Radiation Effects in Silicon Carbide

by
A.A. Lebedev

The book reviews the most interesting, in the author's opinion, publications concerned with radiation defects formed in 6H-, 4H-, and 3C-SiC under irradiation with electrons, neutrons, and some kinds of ions. At the beginning, the SiC electrical parameters making this material promising for application in modern electronics are discussed. Specific features of the crystal structure of SiC are also considered. It is shown that, when wide-bandgap semiconductors are studied, it is necessary to take into account the temperature dependence of the carrier removal rate (η_e), which is a standard parameter for determining the radiation hardness of semiconductors. The η_e values obtained by irradiation of various SiC polytypes with n- and p-type of conductivity are analyzed in relation to the type and energy of irradiating particles. The possible physical mechanisms of compensation of the given material are considered. The influence exerted by the energy of charged particles on how radiation defects are formed and conductivity is compensated in semiconductors under irradiation is analyzed.

Further, the possibility to produce controlled transformation of silicon carbide polytype is considered. The involvement of radiation defects in radiative and nonradiative recombination processes in SiC is analyzed.

Data are also presented regarding the degradation of particular SiC electronic devices under the influence of radiation and a conclusion is made regarding the radiation resistance of SiC. Lastly, the radiation hardness of devices based on silicon and silicon carbide are compared.

Radiation Effects in Silicon Carbide

By

A.A. Lebedev

Ioffe Institute

Polytekhnicheskaya 26, St.Petersburg, 194021 Russia

Shura.lebe@mail.ioffe.ru

Published by **Materials Research Forum LLC**
Millersville, PA 17551, USA

Published as part of the book series
Materials Research Foundations
Volume 6 (2017)
ISSN 2471-8890 (Print)
ISSN 2471-8904 (Online)

Print ISBN 978-1-945291-10-4
ePDF ISBN 978-1-945291-11-1

Distributed worldwide by

Materials Research Forum LLC
105 Springdale Lane
Millersville, PA 17551
USA
http://www.mrforum.com

Manufactured in the United State of America
10 9 8 7 6 5 4 3 2 1

Table of Contents

Foreword

Modern civilization needs increasingly more consumable-energy sources in order to sustain progress in society. Atomic energy and the solar-radiation conversion using ground-based and orbital converters will probably be the main energy sources in the future. Efforts to improve the reliability of both atomic power plants and space-technology systems should be based on the use of radiation-resistant electronics. Radiation resistance typically means that the semiconductor or semiconductor-device parameters are not affected by exposure to nuclear radiation: the higher the radiation dose corresponding to the onset of variation in parameters, the higher the radiation resistance. Semiconductors with a high bonding energy, such as diamond, BN, and SiC, are traditionally thought of as radiation-resistant materials. The advances in technology achieved in the last 10–15 years have made it possible to develop SiC based devices that have fulfilled the expected high potential of SiC in respect to switching power and high operation temperatures. It is now of current practical interest to check to what extent the radiation resistance (radiation hardness) of SiC corresponds to theoretical predictions.

The first studies concerned with radiation defects in silicon carbide, was carried out in the 1950s–1960s, confirmed the high radiation hardness of this material [1]. It should be noted that the crystals studied in those years were heavily doped and had a high density of structural defects. As increasingly perfect and pure SiC samples were obtained, their experimentally measured radiation hardness gradually decreased. Reports even appeared in which it was stated that SiC not only fails to surpass silicon in radiation hardness, but is even inferior to the latter in a number of parameters [2–5]. In this context, the aim of this review is to summarize the available experimental data and assess the correspondence between the parameters of SiC-based devices.

A.A. Lebedev

Ioffe Institute, Polytekhnicheskaya 26, St.Petersburg, 194021 Russia

CHAPTER 1

Physical properties of SiC

Abstract

In addition to possessing unique electrical properties, silicon carbide (SiC) can crystallize in different modifications (polytypes). Having the same chemical nature, SiC polytypes may significantly differ in their electrical parameters. In recent years, the world's interest in fabrication and study of heteropolytype structures based on silicon carbide has considerably increased. This chapter considers studies concerned with different SiC polytypes and their electrical parameters. The latest results obtained in the development of high-power devices based on wide-gap semiconductors are examined. It is shown that at present silicon carbide remains the most promising material for high-temperature, radiation-resistant, high-power electronics. Certain factors involving a wide commercial adoption of SiC-based devices are examined.

Keywords

SiC Polytypism, Hexagonality, Power Devices, Wide Bandgap Semiconductors, Figures of Merit

Contents

1.1 Technology development and history for obtaining silicon carbide and fabricating devices on its basis

The chemical compound of silicon with carbon was first discovered in 1824 by Berzelius. In 1849, this compound was obtained by Despretz via reduction of silica with carbon. Independently, SiC crystals were found by Henri Moissan in a study of meteorites in the

1

Diablo Canyon in the Arizona desert. In 1905, the mineral was named moissanite for its discoverer. At present, this designation is used for jewelry made of silicon carbide. In the end of the last century, Acheson suggested and patented a method for industrial manufacture of SiC [6]. The crystals grown by this method were heavily doped (up to 10^{21} cm^{-3}), had no polytypic homogeneity, and were small in size ($10\times10\times3$ mm^3) [7].

While studying samples of this kind, N.J. Round observed in 1907 an emission of light due to the flow of electric current through the crystal [8]. The electroluminescence from silicon carbide was studied in more detail by O.V. Losev in 1923-1940 [9]. He found that one of these kinds of emission is associated with the presence of a particular "active layer" on the surface of the crystal. Later, he demonstrated that this layer has a n-type conductivity, whereas the bulk of the sample is p-type. Losev also revealed the relationship between the rectification and the electroluminescence. Thus, two phenomena most important for semiconductor electronics, electroluminescence and rectifying properties of p-n structures, were first observed in SiC crystals. Unfortunately, the electronic industry was based at that time on vacuum tubes and these discoveries remained unnoticed for several years.

It is known that the industrial interest in semiconductors appeared after W. Shockley discovered in 1949 the transistor effect in Ge crystals. Approximately at that same time J. Lely suggested a new method for the growth of SiC crystals [10]. In this method, single crystals were grown via sublimation as a result of SiC rectification via the vapor phase from hotter to cooler regions. The rectification is performed in an inert medium (argon) at temperatures of 2500-2650 °C. This technique could be used to grow crystals with areas of up to 4 cm^2 with uncompensated impurity concentrations of 10^{16}-10^{19} cm^{-3}. Disadvantages of this method are the high growth temperatures, the uncontrollable nucleation and growth of the crystals.

In the first half of the 1950s, a search was commenced for semiconductor materials capable of working at higher temperatures as compared to Ge. Therefore, numerous researchers were focused on silicon and silicon carbide. In the following 10-15 years, extensive studies were carried out concerned with the properties of SiC and aimed to develop semiconductor devices on its basis.

However, the industrial interest in silicon carbide waned in the early 1970s. Apparently, this was due to the advances in the Si and GaAs technology, incomparable with those for SiC. The significant technological difficulties in growth of silicon carbide and fabrication of devices on is basis resulted in that the parameters of the devices obtained were far from the theoretical expectations. During the subsequent 15 years, studies of SiC properties were continued by several research teams, most of which worked in the former Soviet Union.

In 1970, the sublimation "sandwich method" was suggested for growth of SiC epitaxial layers [11]. The method consisted in that the growth process occurred on bringing together the vapor source and the substrate. The epitaxy was performed in a vacuum, which made it possible to lower the process temperature to 1810-1910 °C.

At the end of the 1970s, a method was also suggested for growth of bulk SiC crystals [12], the modified Lely method. This technique is based on the vapor condensation on a single-crystal substrate. The growth is performed at temperatures of about 2000 °C. The diameter of the ingot being grown is determined by the substrate size (at present, up to 150 mm) [13].

Finally, the interest in silicon carbide as a promising material for semiconductor electronics returned after S Nishino developed the method of chemical vapor deposition (CVD) for epitaxy of 3C-SiC films on silicon substrates [14]. The use of the standard technological equipment and large-area substrates opened-up prospects for fast commercialization of the results obtained. Soon, several types of field-effect transistors were developed on the basis of films of this kind [15]. However, the parameters of these devices and the quality of the films remained poor. In addition, 3C-SiC is the narrowest gap material among silicon carbide polytypes (Eg = 2.4 eV) and surpasses GaP only slightly as regards to the maximum working temperatures. Therefore, 6H-SiC films were soon obtained on 6H-SiC substrates by the same CVD method [16]. Using the technological combination of a modified Lely method (substrate) and CVD epitaxy (epitaxial layers on Si (0001) face) made it possible to produce from silicon carbide nearly all types of semiconductor devices: blue light-emitting diodes (LEDs), UV photodetectors, Schottky diodes, rectifier diodes, field-effect transistors, bipolar transistors and thyristors [17-18].

1.2   Polytypism in silicon carbide

The existence of different crystalline modifications of SiC was discovered in 1912 [19]. Later, this phenomenon was named polytypism, i.e., polymorphism in one direction [20, 21]. Silicon carbide is a prominent representative of polytypic compounds. Strictly speaking, the term 'polytypism' was specially introduced for carborundum, because different crystalline forms of SiC are very close structurally. At present, more than 200 crystalline modifications of SiC are known [22]. All the known polytypes of silicon carbide crystallize in accordance with the laws of close spherical packing to give binary structures constituted by identical layers. These structures differ both in the order in which cubic C and hexagonal H layers are arranged and in the number of such layers in a unit cell.

Figure 1.1 shows schematically the positions of atom centres for a close spherical packing. If the centres of all the spheres lie in the first layer at points A, the positions of centres at points B or C are possible in the second layer. If the centres of all the spheres lie in the layers at points B, the positions of centres at points A or C are possible in the next layer.

Table 1.1. Selected parameters of SiC [31-33]

Parameter	2H-SiC	4H-SiC	6H-SiC	3C-SiC
Stacking order	AB	ABCB	ABCACB	ABC
Jagodzinskii notation	h	hc	hcc	c
"Hexagonality" percentage, γ, %	100	50	33	0
Lattice constant, Å	a = 3,076 c = 5,048	a = 3,073 c = 10,053	a = 3,08 c = 15,117	4,34
Band gap, eV	3,33	3,26	3,0	2,39
Thermal conductivity, (W/cm· deg)		3-4	3-4	3-4
Critical breakdown field strength, E_{cr}, (MV/cm)		2-3	2-3	>1,5
Electron mobility μ_n, cm^2/(V·s)(300 K)		≤ 850	≤ 450	≤1000
Saturation velocity V_s, 10^7 cm/s		2	2	2,7
Hole mobility μ_p, cm^2/(V·s)(300 K)		≤ 120	≤ 100	≤ 40

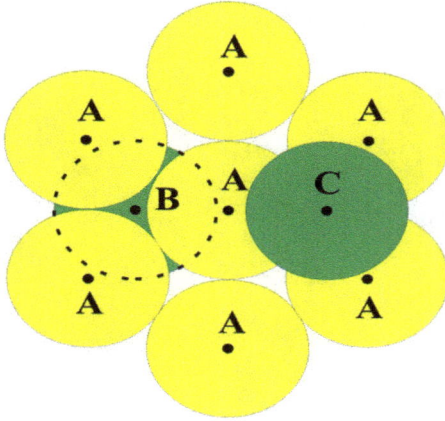

Figure 1.1. *Schematic positions of atom centres for a close spherical packing. There are only three possible positions for atom centres—A, B and C.*

If centres of all the spheres lie at points C in a layer, the positions of centres at points A or B are possible in the next layer. In such a way, the positions occupied by atoms in the second and subsequent layers determine the structure of a polytype. In other words, for each polytype, the structure will simply correspond to the letters' sequence AB, ABC, ABCB, etc. (Table 1.1). Polytypes are also frequently characterized by Ramsdell designations [35], constituted by a natural number equal to the number of layers in the period in the direction perpendicular to the basal plane and a letter symbol characterizing the crystal system of the Bravais lattice: C, cubic; H, hexagonal; R, rhombohedral. The most frequently occurring are 6H, 4H, 15R and 3C polytypes (figures 1.2 and 1.3). Although the positions of nearest-neighbour atoms are the same for any silicon or carbon atom in all the polytypes, the positions of farther neighbours differ, which leads to the existence of crystallographically nonequivalent positions in the SiC lattice. Only in two SiC polytypes, the positions of all atoms are equivalent and correspond to either cubic (3C-SiC) or hexagonal (2H-SiC) sites of the crystal lattice (4H, one cubic and one hexagonal; 6H, two cubic and one hexagonal) [24]. In all other polytypes, atoms may occupy sites of both types and polytypes differ in the number of atoms in hexagonal (NH) and cubic (NC) positions (see, e.g. [25]). Therefore, it is convenient to characterize silicon carbide polytypes by the parameter 'hexagonality' γ [26], which is defined as the ratio between the number of atoms in hexagonal positions and the total number of atoms in the unit cell:

$$\gamma = NH/(NH + NK).$$

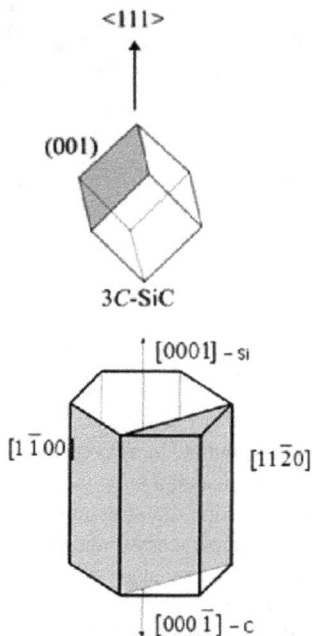

Figure 1.2. *Schematic images of the hexagonal and cubic crystal lattices of SiC. Adapted from [35].*

The hexagonality of a polytype may vary from unity (2H-SiC) to zero (3C-SiC) (table 1). It is noteworthy that the ability to crystallize in different crystal lattices is inherent not only in SiC, but also in quite a number of other compounds: GaN, ZnSe, ZnO, diamond, etc.

It is known that silicon carbide is a semiconductor with a nondirect band structure. The band gap width strongly depends on the polytype and varies from 2.39 eV for 3C-SiC to 3.3 eV for 2H-SiC (figure 1.4). Thus, the difference in the band gap widths ($Eg_{[H-C]}$) of cubic (3C) and purely hexagonal (2H) polytypes of SiC is ~ 0.9 eV. It should be noted that, for other semiconductors having polytypes (GaN, ZnS, ZnSe, etc.), $Eg_{[H-C]} \leq 0.2$ eV [27]. In addition, these compounds, as a rule, crystallize only in the 3C and 2H forms, with no intermediate polytypes existing. By contrast, the extreme polytypes could not be obtained for SiC, whereas a multitude of intermediate forms exist for this compound. 3C-SiC films of a rather large area have been obtained [28], whereas 2H crystals presently exist only in the form of needles [29]. The absence of crystals with sizes suitable for device applications result in that most of the electrical parameters of 2H-SiC have not been determined. According to the results of experimental and theoretical studies, the valence-band maximum is at the centre of the Brillouin zone and the conduction-band minimum lies at its boundary. To this fact is attributed the strong dependence of the band gap width on the structure of a polytype. According to [30], the spin–orbit splitting of the valence band is 10 meV. The main parameters of 6H, 4H and 3C SiC polytypes are listed in table 1.1.

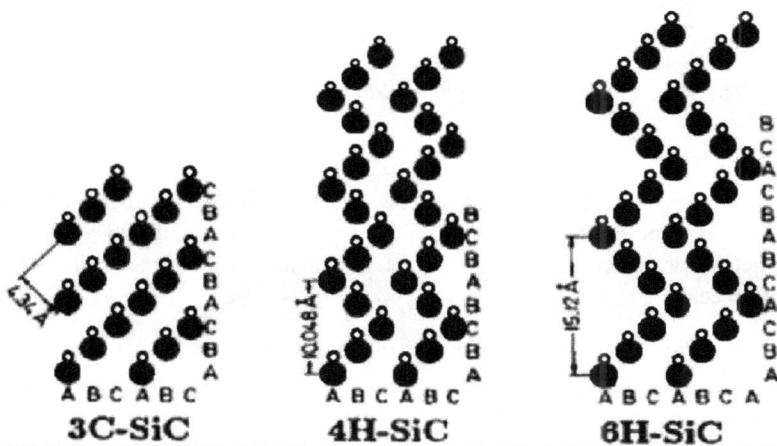

Figure 1.3. *Positions of Si and C atoms (open and full circles, respectively) in the (11 20) plane for different SiC polytypes. Adapted from [31].*

7

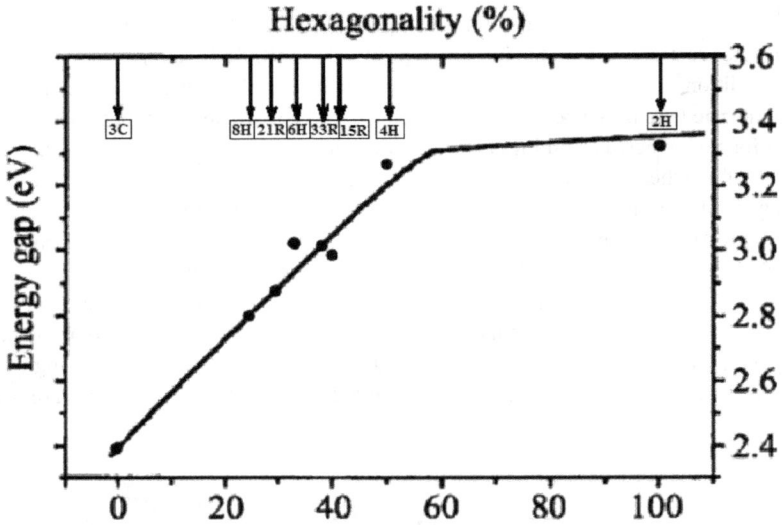

Figure 1.4. Exciton energy gap of different SiC polytypes at 300 K as a function of the hexagonality [36].

At present, there is no theory that would be satisfactory in every respect in explaining why SiC and some other materials crystallize in a wide variety of polytypes. It is not completely clear, either, what factors favour formation of one or another polytype [34 and references in it].

1.3   SiC parameters important for electronics

In modern semiconductor physics, there are now two very rapidly developing directions of research:

1. changing the properties of a material by modifying the geometric dimensions of the structures, i.e., the physics of nanostructures, and

2. the development and study of the new semiconductor materials.

Important work in the second direction involves the study of wide-bandgap semiconductors. The potentialities of wide-bandgap materials for creating semiconductor devices have been well analyzed [37-39]. A band gap larger than that in Si and GaAs gives these materials the following advantages:

- a high operating temperature;
- ability to construct visible-range light-emitting devices based on these materials;
- high critical breakdown fields (E_{cr});
- high radiation resistance.

Table 1.2 lists the main parameters of some semiconductor materials. The Debye temperature (T_D) is presented among other properties of semiconductor materials. It is known that this parameter is defined in terms of the maximum energy of a phonon (E_f) existing in a given material

$$T_D = E_f/k$$

where k is the Boltzmann constant.

As the Debye temperature is exceeded, lattice vibrations become inelastic and this leads to disintegration of the material. For various semiconductors, this disintegration may occur via growth of dislocations, decomposition of binary compounds, etc. Thus, the value of T_D can be regarded as the upper temperature limit for operation of devices based on a given material. It is noteworthy that the T_D of GaN is lower than that of SiC. There also exists a whole class of wide-bandgap materials (II - VI) for which T_D is even lower than that of GaN. This circumstance accounts for the fact that it has been impossible so far to fabricate any high-power high-temperature devices based on II - VI compounds, despite their wider energy gap.

Table 1.2 also indicates whether a given semiconductor is a direct- or indirect-gap semiconductor and whether a large-diameter substrate consisting of the same material is present (+) or absent (-). The first factor largely determines the applicability of a given material for optoelectronic devices. The second factor determines the application possibilities of the scientific results in a commercial production.

It is known that special quality criteria, figures of merit (f_m) calculated using the basic physicochemical properties of a semiconductor, have been proposed for comprehensively accounting for the potentialities of a semiconductor material.

Johnson [37] proposed to use the product of E_{cr} and the saturation velocity V_{sat} as a figure of merit JM, which determines the operating limit of a conventional transistor: JM = $(E_{cr}V_{sat}/\pi)^2$. Later, the following criterion was defined: KM= $\lambda(V_{sat}/\epsilon)^{1/2}$; ($\epsilon$ is the dielectric constant of the semiconductor, and λ, thermal conductivity), where the switching rate of a transistor operating as a logic element of a computer was taken into account [40]. Baliga [41] proposed yet another figure of merit BM for assessing a semiconductor material. It is related to the operating loss of high-power field-effect transistors: BM = $\mu \in E_{cr}^3 \sim (\mu$ is the carrier mobility).

Table 1.2. Parameters of some semiconductor materials.

	Si	GaAs	4H-SiC	GaN	Diamond
Eg, eV	1,12	1,43	3,26	3,39	5,45
Direct band structure		+		+	
Breakdown field, MV/cm	0,3	0,6	3	>3	10
Thermal conductivity, Wt/cm*K	1,5	0,46	4,9	1,3	11
Electron mobility, cm$^{2/}$V*s	1500	8500	800	1250	2200
Presence of a native substrate	+	+	+	-	-
Debay temperature	650	350	1200	600	1850
Saturation velocity, x 10^7 cm/s	1	1	2	2,5	2,7
Dielectric constant, ϵ	11,7	12,9	10	10	5,7

However, this criterion was primarily associated with the ohmic loss and used to assess the performance capabilities of a semiconductor from the standpoint of low-frequency devices. For high-frequency devices, the switching loss should also be examined. The criterion BH = μE_{cr}^2, based on the assumption that the switching loss is due to the recharging of the input capacitance of a device, was proposed in [42]. Other important properties to be analyzed when selecting materials for high-power devices are the thermal properties of a semiconductor. A figure of merit that takes into account these properties was proposed in [43]: QF = $\lambda\mu E_{cr}^3$.

The computed relative values of five figure-of-merit parameters for 4*H*-SiC, Si, GaAs, GaN, and diamond are listed in Table 1.3. In the calculation, the values of the parameters for Si were taken as the measurement unit.

Table 1.3 Normalized figures of merit for certain semiconductors

	$JM = (E_{cr}V_{sat}/\pi)^2$	$KM = \lambda(V_{sat}/\epsilon)^{1/2}$	$BH = \mu E_{cr}^2$	$BM = \mu\epsilon E_{cr}^3$	$QF = \lambda\mu E_{cr}^3$
Si	1	1	1	1	1
GaAs	4	0,3	23	50	15
4H-SiC	400	5	53	456	1489
GaN	624	1,5	83	712	617
Diamond	8100	17	1630	26455	194000

It can be seen in Table 1.3 that silicon carbide surpasses, nearly in all the criteria considered above, the classical semiconductor materials, Si and GaAs. It is of interest to compare SiC with other wide-bandgap materials.

SiC is inferior to GaN in a number of parameters. For example, the low probability of radiative recombination (as in an indirect-gap semiconductor). However, native substrates are still unavailable for GaN, whereas for AlN, native substrates are very small in size and exceedingly expensive. GaN is grown by heteroepitaxy on substrates of other materials (SiC, sapphire). This leads to a very high dislocation density in the films being grown ($>10^7$ cm^{-2}). The dislocations in GaN are oriented perpendicularly to the surface of a growing layer and aggregate into clusters. As a result, the growing layer has a cellular (grainy) structure, which increases the leakage current of the pn structures and causes their degradation in the course of time.

On the whole, SiC is the more promising material for device fabrication, compared with GaN and other nitrides (III-N). The unattainable (because of the higher probability of radiative recombination) longer carrier lifetime in GaN restricts the applicability of this material in the development of bipolar devices. In unipolar devices, its low heat conductivity and the lower Debye temperature reduce the maximum dissipated power. Bulk GaN has no significant advantages over silicon carbide, either, (in other electrical parameters (saturation velocity of carriers, breakdown field, carrier mobility). However, it is possible that, owing to their substantially lower production cost, GaN Schottky diodes will be competitive with SiC Schottky diodes at voltages of up to 1000 V [44].

However, the development of heterojunctions in the GaN - AlGaN system made it possible to obtain structures with a 2D electron gas having a substantially higher carrier mobility. These structures can be used to develop high-frequency (high-electron-mobility transistors, HEMTs) whose parameters exceed those of SiC-based field-effect transistors.

Diamond is beyond competition both in its parameters and in the maximum working temperatures. However, single-crystal diamond layers still cannot be produced by heteroepitaxy and native substrates have a small area and are rather expensive. Moreover, there are certain difficulties in obtaining pn structures for diamond.

Thus, the final conclusion can be formulated as follows: regarding the fabrication of power devices, SiC is presently the most promising material among those best developed and the best developed among those most promising.

The parameters of some commercially available devices based on silicon carbide are presented in Table 1.4

Table 1.4 SiC device parameters

Devices type	Continuous forward current, A	Reverse Voltage (Blocking voltage), V	Operating Temperature, ^{0}C	Switching frequency, kHz	Drain-Source On-State Resistance, mΩ	DC current Gain
Schotky diodes [45]	15	1200	-55....+175			
JFET [46]		1200		48		
MOSFT [47]	42	1200			80	
BJT [48]	10-50	1200				52-45

References

[1] W. J. Choyke, A review of radiation damage in SiC, Inst. Phys.: Conf. Ser. 31, 58 (1977).

[2] A. Hallen, A. Henry, P. Pelligrino, B. G. Swensson, and D. Aberg, Lateral spread of implanted ion distributions in 6H-SiC: simulation, Mater. Sci. Eng. B 61–62, (1999) 378-381. http://dx.doi.org/10.1016/S0921-5107(98)00538-8

[3] B. G. Svensson, AQ.Hallen, M.K.Linnarson, A.Yu.Kuznetsov, M.S.Janson, D.Aberg, J.Osterman, P.O.A. Persson, L.Hultman, L.Storasta, F.H.C. Carlsson, J.P.Bergman, C.Jagadish and E.Morvan, Mater. Sci. Forum 353–356, (2001) 549-552. http://dx.doi.org/10.4028/www.scientific.net/MSF.353-356.549

[4]   G. Casse, Overview of the recent activities of the RD50 collaboration on radiation hardening of semiconductor detectors for the sLHC, Nucl. Instrum. Methods Phys. Res. A 598, (2009) 54-60. http://dx.doi.org/10.1016/j.nima.2008.08.019

[5]   J. Metcalfe, Silicon Detectors for the sLHC, Nucl. Phys. B (Proc. Suppl.) 215, (2011) 151-153. http://dx.doi.org/10.1016/j.nuclphysbps.2011.03.162

[6]   E.G. Acheson. "On carborundum" Chem. News, 68, 179 (1893).

[7]   G. Pensl, R. Helbig. In: *Advances in Solid State Physics*, ed. by V. Rossler (Viemeg, Braunschweig, 1990).

[8]   N.J. Round, A note on carborundum, Electrical World, 30, 309 (1907).

[9]   O.V.Losev, The photoelectric effect is the in a special active layer of carborund crystal Sov.Tech.Phys 1, 718 (1931).

[10]   J.A. Lely,Darstellung von Einkristallen von Siliziumcarbid und Beherrshing von Art und mende der eingebeuten verunreininungen, Ber. Dt. Keram. Ges., 32, 229 (1955).

[11]   Yu.A. Vodakov, E.N. Mokhov, M.G. Ramm, A.O. Roenkov, Epitaxial growth of silicon carbide layers by sublimation "sandwich-method", Krist and Tecnik., 14, 729 (1979). http://dx.doi.org/10.1002/crat.19790140618

[12]   Yu.M. Tairov, V.F. Tsvetkov, Investigation of Growth procces of ingots of silicon carbide single crystals, J. Cryst. Growth, 43, 209 (1978). http://dx.doi.org/10.1016/0022-0248(78)90169-0

[13]   D.Chaussende, K.Ariyawong, N.Tsavdaris, M.Seiss, Y.J. Shin, J-M. Dedulle, R.Madar, E. Sarigiannidou, J.La Manna, O. Chaix-Pluchery, T.Quisse, Open issues in SiC bulk growth, Mat.Science Forum 778-780 (2014) pp 3-8. http://dx.doi.org/10.4028/www.scientific.net/MSF.778-780.3

[14]   S. Nishino, J. Powell, N.A. Will. Production of large-area single-crystal wafers of cubic SiC for semiconductor devices, Appl. Phys. Lett., 42, (1983) 460-462. http://dx.doi.org/10.1063/1.93970

[15]   H.S. Kong, J.W. Palmer, J.T. Glass and R.F.Davis, Temperature dependence of the current-voltage characteristics of metal-semiconductor field-effect transistors in *n*-type β-SiC grown via chemical vapor deposition, Appl Phys Lett, 51, (1987) 442-444. http://dx.doi.org/10.1063/1.98416

[16]   J.W. Palmer, H.S. Kong, R.F. Davis. High-temperature depletion-mode metal-oxide-semiconductor field-effect transistors in beta-SiC thin films, Appl. Phys. Lett., 51, (1987) 2028-2030. http://dx.doi.org/10.1063/1.98282

[17]   J.A. Edmond, H.S. Kong, C.H. Carter. Blue LEDs, UV photodiodes and high-temperature rectifiers in 6H-SiC, Physica B, 185, (1993) 453-460. http://dx.doi.org/10.1016/0921-4526(93)90277-D

[18]   J.W. Palmer, J.A. Edmond, H.S. Kong, C.H. Carter, 6H-silicon carbide devices and applications, Physica B, 185, (1993) 461-465. http://dx.doi.org/10.1016/0921-4526(93)90278-E

[19]   Baumhauer H 1912 Uber die Krystalle des Carborundums Z. Kristallogr. 50 33

[20]   Baumhauer H 1915–1920 Uber die Modificationen des Carborundums und die erscheinung der Polytypie Z. Kristallogr. 55 249

[21]   Schneer C J Polytypism in one dimension Acta Crystallogr. 8 (1955) 279. http://dx.doi.org/10.1107/S0365110X55000893

[22]   Verma A R and Krishna P 1966 Polymorphism and Polytypism in Crystals (New York: Wiley)

[23]   Ramsdell L S 1947 Studies on silicon carbide Am. Mineral. 32 64

[24]   Patric L 1962 Nonequivalent sites and multiple donor and acceptor levels in SiC polytypes Phys. Rev. 127 1878. http://dx.doi.org/10.1103/PhysRev.127.1878

[25]   Henisch H K and Roy R (ed) 1968 Silicon Carbide-1968 (Oxford: Pergamon)

[26]   Jagodzinskii H 1949 Polytypism in SiC crystals Acta Crystallogr. 2 201

[27]   Steckl A L and Zavada J M 1999 Optoelectronic properties and applications of rare-earth-doped GaN MRS Bull. 24 33. http://dx.doi.org/10.1557/S0883769400053045

[28]   Nagasaws H, Yagi K, Kawahara T and Hatta N 2003 Properties of free-standing 3C-SiC monocrystals grown on undulant-Si(0 0 0 1) substrate Mater. Sci. Forum 433–436

[29]   Dudley M, Huang W, Vetter W M, Neudeck P and Powell J A 2000 Synchrotron white beam topography studies of 2H SiC crystals Mater. Sci. Forum 338–342 465. http://dx.doi.org/10.4028/www.scientific.net/MSF.338-342.465

[30]   Humphreys R G, Bimberg D and Choyke W J 1981 Wavelength-modulated absorption in SiC Solid State Commun. 39 163. http://dx.doi.org/10.1016/0038-1098(81)91070-X

[31] Matsunami H and Kimoto T Step-controlled epitaxial growth of SiC: High quality homoepitaxy, 1997 *Mater. Sci. Eng.* R 20 125. http://dx.doi.org/10.1016/S0927-796X(97)00005-3

[32] W von Munch 1982 Silicon carbide Landolt-Bounstien: Numerical Data and Functional Relationships in Science and Technology vol 17A (Berlin: Springer) p 132

[33] Levinshtein M E, Rumyantsev S L and Shur M S 2001 Properties of Advanced Semiconductor Materials: GaN, AlN, InN, BN, SiC, SiGa (New York: Wiley)

[34] A.A.Lebedev, Hetrojunctions and superlattices based on silicon carbide. Topical review, Semiconductor Science Technology 21 (2006) R17-34. http://dx.doi.org/10.1088/0268-1242/21/6/R01

[35] Takagi H, Nishiguchi T, Ohta S, Furusho T, Ohshima S and Nishino S 2004 Crystal growth of 6H-SiC (0 1 ‾14) on 3C-SiC (0 0 1) substrate by sublimation epitaxy *Mater. Sci. Eng.* 457–460 289

[36] A. Fissel Artificially layered heteropolytypic structures based on SiC polytypes: molecular beam epitaxy, characterization and properties, Phys. Rep. 379 (2003) 149. http://dx.doi.org/10.1016/S0370-1573(02)00632-4

[37] E.O. Jonson, Physical limitations on frequency and power parameters of transistors", *RCA Review*, 26, (1965) 163-177. http://dx.doi.org/10.1109/irecon.1965.1147520

[38] B.J. Baliga, Semiconductors for high-voltage, vertical channel field transistors, J. Appl. Phys, 53, (1982) 1759-1764. http://dx.doi.org/10.1063/1.331646

[39] A.A. Lebedev and V.E. Chelnokov, Wide-gap semiconductors for high-power electronics, *Semiconductors*, 33, (1999) 999-1001. http://dx.doi.org/10.1134/1.1187823

[40] R. W. Keyes, Figure of merit for semiconductors for high-speed switches, Proc. IEEE 60, 225 (1972). http://dx.doi.org/10.1109/PROC.1972.8593

[41] B. J. Baliga, Semiconductors for high-voltage, vertical channel field-effect transistors J. Appl. Phys. 53, 1759 (1982). http://dx.doi.org/10.1063/1.331646

[42] R. Gaska, Q. Chen, J. Yang, A. Osinsky, M. Asif Khan, and M. S. Shur, High-temperature performance of AlGaN/GaN HFETs on SiC substrates,IEEE Electron Device Lett. 18, 492 (1997). http://dx.doi.org/10.1109/55.624930

[43] K. Szenai, R. S. Scott, and B. J. Baliga, *Optimum semiconductors for high-power electronics,*I EEE Electron. Dev. 10, 85(1989).

[44] Lei Yong, Shi Hongbiao, Lu Hai, Chen Dunjun, Zhang Rong and Zheng Youdou. Field plate engineering for GaN – based A Schottky barrier diodes, *J. of semiconductors,* 34 (2013) 054007.

[45] SiC Schottky diode IDW15S120 datasheet, Infineon

[46] Ralf Siemieniec and Uwe Kirchner, "The 1200V Direct-Driven SiC JFET power switch," EPE 2011

[47] SiC MOSFET, CMF20120D datasheet, Cree

[48] SiC BJT FSICBH017A120 datasheet, Fairchild Semiconductor

CHAPTER 2

Compensation of silicon carbide under irradiation

Abstract

It is shown that for the investigations of wide-gap semiconductors (WGS) account should be taken of how the rate of removal of mobile charge carriers—the standard parameter in determining the radiation hardness of a material—depends on the temperature. The use of data obtained only at room temperature may lead to an incorrect assessment of the radiation hardness of WGS. A conclusion is made that the WGS properties combine, on the one hand, high radiation hardness of high-temperature devices based on these semiconductors and, on the other, the possibility of effective radiation-induced doping (e.g., for obtaining semi-insulating local regions in a material at room temperature). The model of conductivity compensation in SiC under irradiation is presented. Processes of radiation defect formation and conductivity compensation in silicon and silicon carbide irradiated with 0.9 MeV electrons are considered in comparison with the electron irradiation at higher energies.

Keywords

Charge Carriers Removal Rate, Threshold Energy, Radiation Hardness, Radiation Doping, Conductivity Compensation, Primary and Secondary Radiation Defects, Recoil Atoms.

Contents

2.1   Threshold energy of defect formation.

As already noted, the radiation resistance is, as a rule, better in semiconductors with a higher bonding energy. In order to characterize this relationship, a parameter is introduced, the threshold defect formation energy Ed. This parameter represents the minimal energy to be transferred by a particle to the semiconductor matrix to form a Frenkel pair, i.e., a vacancy and an interstitial atom [1, 2]. A theoretical calculation of the value of Ed involves a solution of a many-body problem and encounters a number of difficulties associated with the choice of the type and constants of the interaction potentials and other parameters [2]. To experimentally determine Ed, the variation of only a single chosen parameter under the effect of irradiation is usually monitored, although radiation affects all properties of a semiconductor simultaneously. As a result, the values of Ed exhibit a wide scatter and depend on the experimental procedure used in a study. In [3], a relationship between the value of Ed and the lattice constant a_0 of a given semiconductor was observed, which satisfies the following phenomenological relation:

$$1,117 Ed = (10/a_0)^{4.363} \qquad (2.1)$$

Table 2.1 lists values of Ed calculated by the above formula (2.1) for a number of semiconductors and the experimental data obtained when determining Ed. It can be seen in the table that there is a significant scatter of the Ed values experimentally found for silicon carbide. Possibly, this is due to the poor structural quality of SiC crystals, especially in studies carried out before the mid-1990s. The value of Ed was found to be larger for cubic silicon carbide (3C-SiC) and smaller for hexagonal silicon carbide, compared with the values calculated by formula (2.1). Because the structural quality of 3C-SiC crystals is still substantially poorer than that of 6H- and 4H-SiC, it can be assumed that Ed does not exceed 30-35 eV for all SiC polytypes. This value of Ed is 1.5-2 times larger than that for Si and GaAs and is 2-2.5 times smaller than that for a diamond. However, additional experiments are necessary in any case for determining the exact value of Ed for present-day high-quality epitaxial SiC films. It is also important to determine the temperature dependence of Ed for SiC as a material that is promising for application in high-temperature electronics.

The values of threshold energies make it possible to calculate the number of primarily created radiation defects. The data obtained in [8] for materials that are of interest for high-energy physics (detection of relativistic particles and cosmic-ray fields) are listed in Table 2.1. and 2.2. It can be seen that SiC is only slightly inferior to diamonds in this parameter, but noticeably surpasses silicon and, even more so, gallium arsenide.

Table 2.1. Calculated and experimental values of the threshold defect-formation energy for some semiconductors[5].

	GaAs	Si	3C-SiC	Diamond	6H and 4H SiC
Lattice constant a_0	5,65	5,431	4,36	3,57	3,08
Calculated value of Ed (formula 2.1), eV	9	12,8	37	80	153
Experimental value of Ed, eV.	8-20 [2]	13-20 [2]	106 [6] 54-90[7]	60-80 [2]	97 [4] 20-35 [5]

Table 2.2. Number of appearing primary radiation defects for a number of semiconductors, normalized to the corresponding value for silicon carbide [9].

Semiconductors	Protons	Pions	Cosmic rays
Diamond	1,2	0,7	0,5
Silicon	2,4	4,1	3,3
Gallium Arsenide	12	43,8	27,5

2.2   Temperature dependence of the carrier removal rate

Radiation hardness is commonly understood as the stability of parameters of a semiconductor or a semiconductor device under nuclear irradiation. The higher the irradiation dose that is necessary for the parameters to start changing, the better the radiation hardness of a semiconductor.

It is known that irradiation of semiconductors results in deep centers - radiation defects (RDs) of acceptor or donor nature - being formed. Henceforth, we consider the irradiation of an n-type material for the sake of definiteness. In this case, electrons transit under irradiation from the conduction band to deep RDs of acceptor nature. As a result, the conductivity of the material decreases, and the semiconductor may become an insulator at large irradiation doses. This process is commonly described for various materials by using the carrier removal rate η_e (cm^{-1}) as parameter [1].

$$\eta_e = \Delta n/\, D = (n_0 - n\,)/\, D \qquad\qquad (2.2)$$

Here, n_0 and n are, respectively, the carrier densities in the conduction band before and after the irradiation, and D is the irradiation dose.

Let us assume that the Fermi level lies in the energy gap several kT below the main donor levels, and the total concentration of the RDs introduced is substantially lower than Nd - Na. That is, it can be assumed that irradiation does not change significantly the Fermi level position at low irradiation doses (and just at these doses the carrier removal rate is commonly determined). Let us now consider the contribution to η_e from the RDs formed, depending on their type (donor or acceptor) and position in the energy gap. All the RDs being formed can be conditionally divided, depending on their energy position, into three groups.

(1) RDs lying higher than the main donor levels. If these RDs are donors, they will contribute to an increase in the carrier concentration in the band by N_{D1}. If these RDs are acceptors, their formation will have no effect on the carrier concentration in the conduction band. With increasing measurement temperature, the contributions from both types of RDs to the value measured for η_e will remain unchanged.

(2) RDs lying in the upper half of the energy gap below the main donor levels and the Fermi level. If these RDs are acceptors, their introduction will reduce the electron concentration in the conduction band by N_{A2}. If these RDs are donors, they will make no contribution to the free electron concentration. With increasing temperature (or with the Fermi level going down as a result of irradiation), these RDs will find themselves within several kT from the Fermi level. The degree of electron ionization from these RDs will

grow and, accordingly, the contribution from acceptor levels to the value of η_e will decrease and those from the donor levels will increase.

(3) RDs lying in the lower half of the energy gap. If these RDs are acceptors, they will contribute to a decrease in the carrier concentration in the band by N_{A3}. If these RDs are donors, their formation will have no effect on the carrier concentration in the band. With increasing measurement temperature, the contributions from both types of RDs to the value measured for η_e will remain invariable.

Table 2.3 presents the contribution and its temperature dependence for each type of the RDs mentioned above. Henceforth the Fermi level position is reckoned from the valence band top. Accordingly, the following formula can be written for η_e with consideration for the temperature dependence of the RD occupancy:

$$n = n_0 + D\{K_{D1} - K_{A3}/(1+ \lambda^{-1}{}_{D2}\exp[(E_F\text{-}E_{D2}/kT]) - K_{A2}(1 + A_2\exp[(E_{A2} - E_F)/kT])\} \quad (2.3)$$

Here, K_j is the total rate of introduction of jth type of defects, $N_j = K_jD$, λ^{-1} is the degeneracy factor of the levels related to the RDs, k is the Boltzmann constant, and T is absolute temperature.

Formula (2.2) can be written in simplified form as

$$n = n_0 - (\Delta n_1 + \Delta n (T_M)) \quad (2.4)$$

where Δn_1 is the temperature-independent change in the carrier concentration, and $\Delta n(T_M)$ is the change in the carrier concentration, which depends on the measurement temperature T_M.

Accordingly, we can write for the carrier removal rate

$$\eta_e = \eta_{e1} + \eta_e (T_M) \quad (2.5)$$

An important conclusion follows from the formulas (2.2) - (2.5): if two temperatures, $T_{M1} < T_{M2}$, then $\eta_e(T_{M1}) \geq \eta_e(T_{M2})$, i.e., the value of η_e can only decrease with increasing measurement temperature. Therefore, $\eta_e \approx 0$ at the limiting working temperatures ($E_f \approx 1/2Eg$) if only acceptor RDs were formed under irradiation, and $\eta_e(T_M) < 0$ if donor RDs were also generated. If the total concentration of donors formed upon irradiation exceeds that of the resulting acceptors, the overall value of η_e may be negative at high T: $\eta_e + \eta_e(T_M) < 0$. If this is the case, it would be more appropriate to speak about the rate of carrier *introduction* into the semiconductor under irradiation.

Table 2.3. Contribution to the carrier concentration from an RD, depending on its type and energy position in the forbidden gap [10]

Groups of levels	Position of an RD at T_M=Tirradiation	Type of RD	Contribution to the carrier concentrationй (Δn) at T_M= Tirradiation	Change in Δn at T_M> Tirradiation (Δn=f(T)) (with the degeneration factor disregarded)
1	Ef$-E_{RD}$ <0	A	0	none
	Ef$-E_{RD}$ <0	D	$+ N^{DI}$	none
2	Ef$-E_{RD}$ >0; $E_{RD} \geq 1/2$ Eg	A	$- N^{A2}$	$-N_{A2}/[(1+\lambda \cdot \exp((E_{A2}-E_f)/kT)]$
	Ef$-E_{RD}$ >0; $E_{RD} \geq 1/2$ Eg	D	0	$+N_{D2}/[(1+\lambda^{-1} \cdot \exp((E_f-E_{D2})/kT)]$
3	E_{RD}< 1/2 Eg	A	$- N^{A3}$	none
	E_{RD}< 1/2 Eg	D	0	none

These difficulties necessitate a search for other ways to compare radiation effects. For example, there is no need to heat a sample to the limiting working temperatures and measure the carrier concentration n_0 in order to determine η_e in the whole range of irradiation doses. As a measure of n_0 can serve the net concentration $|N^+_D - N^-_A|$ in the space charge region. In this case, it is sufficient to measure capacitance-voltage (C - V) characteristics and this can be done at lower temperatures. Whether or not a particular level contributes to the value of $|Nd - Na|$ measured at a given T depends on the relationship between the time constant of its ionization (τ) and the time t during which C--V characteristic is measured, $t/\tau > 1$, i.e., $t\eta = \tau$ ($\eta < 1$). Each semiconductor is characterized by its own temperature T_{CV} at which the limiting value of η_e can be evaluated by the capacitance method. It is known that

$$\tau^{-1} = V_T \sigma_N Nc \exp(-Ei/kT) \qquad (2.6)$$

where V_T is the thermal velocity of carriers, σ_N is the cross-section of electron capture by a level, N_c is the density of states in the conduction band, E_i is the ionization energy of a center, and k is the Boltzmann constant.

Hence we obtain

$$T_{CV} = Ei/(k \ln(V_T \sigma_N Nc \eta t)) \qquad (2.7)$$

Assuming that $E_i = (1/2)Eg$, $\sigma_N = 10^{-16}$ cm^{-3}, $t = 600$ s, and $\eta = 0.2$, we can estimate T_{CV} for Si and 4H-SiC to be, respectively, ~230 and ~ 630 K. This estimate shows that the C - V method can be used at room temperature to obtain the limiting value of η_e for silicon. This temperature is insufficient for determining η_e for wide-bandgap semiconductors (WBSs).

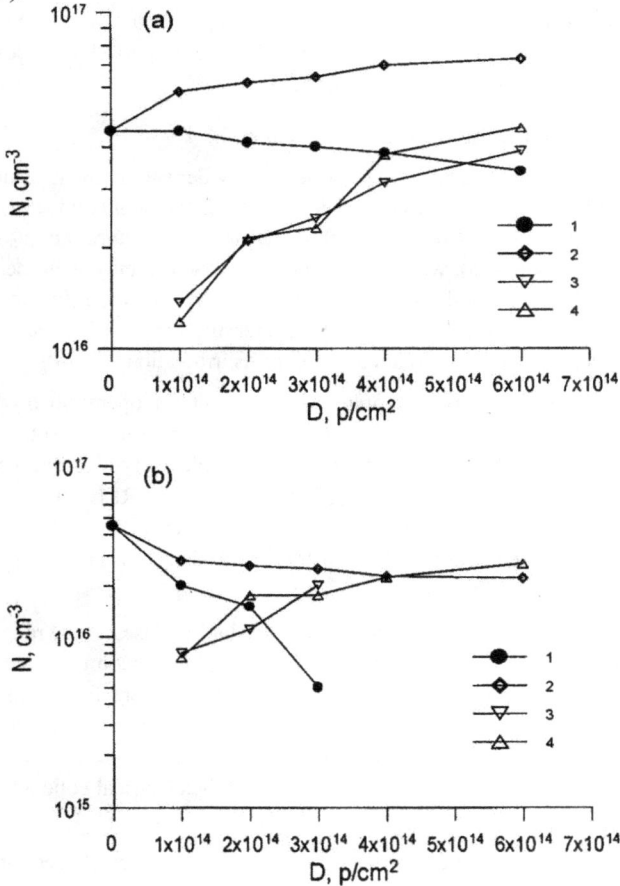

Fig 2.1 (A) 6H-SiC, Nd -- Na at T = 300 K (1) and T = 650 K (2), difference of these (3), and the concentration of the center at Ec - (1.1-1.22 eV (4) vs. the irradiation dose. (B) 4H-SiC, Nd - Na at T = 300 K (1) and T = 650 K (2), difference of these (3), and total concentration of the centers (RD$_{1/2}$ + RD$_3$ + RD$_4$) (4) vs. the irradiation dose [11].

An experimental study of the temperature dependence of η_e in 4H and 6H SiC samples irradiated with 8 MeV protons was reported in [11]. C-V characteristics were measured on a standard C-V installation with a parallel equivalent circuit at a sinusoidal frequency of 10 kHz. A study of irradiated samples demonstrated a decline in the Nd - Na value measured at room temperature, with Nd - Na strongly increasing after the structure was heated to 650 K. For 6H-SiC the value of Nd - Na measured at 650 K was even higher than that in the initial structures prior to irradiation. With increasing irradiation dose, this difference became more pronounced (Figs. 2.1^A and 2.1^B).

2.3 Dependence of η_e on the measurement procedure

It should also be noted that the experimental value of η_e may depend on the measurement technique. For example, presence of deep centers (DCs) in the semiconductor before irradiation will lead to an additional rise in n with increasing temperature, i.e., $n_0 = f(T)$. The situation when $n_0 = \gamma(\text{Nd - Na})$, where $\gamma < 1$ and γ approaches unity with increasing T, is typical of WGSs in general and SiC in particular. To eliminate the influence of the temperature dependence of n_0 on the results of η_e measurements, it is necessary to measure this dependence in a sample under study before its irradiation.

In [12], two electrical parameters were monitored, important for operation of devices using the charge- and current-related characteristics of a p-n junction: concentrations of uncompensated acceptors and free carriers (holes). As samples served p-SiC epitaxial layers were fabricated by sublimation epitaxy [13] on commercial CREE substrates. The thickness of the p-layer was about 300 µm, and that of the substrate, about 400 µm. The concentration of uncompensated donors in the substrate, (Nd - Na), was $(3-5) \times 10^{18}$ cm^{-3}, and that of acceptors in the p-layer, $\sim (1-2) \times 10^{18}$ cm^{-3}.

The irradiation with 8 MeV protons was performed at irradiation doses in the range from 1.0×10^{15} to 1.1×10^{16} cm^{-2} on an MGTs-20 cyclotron. The uncompensated donor concentrations Nd - Na in the as-grown and irradiated samples were found from C-V measurements. The C-V characteristics were measured with a mercury at room temperature. The concentrations of free carriers and their mobilities were determined at the same irradiation doses by measuring the Hall effect and the electrical conductivity by the Van der Pauw method.

The experimentally obtained dependences of Na - Nd and free carrier concentration on the irradiation dose are shown in Fig. 2.2.

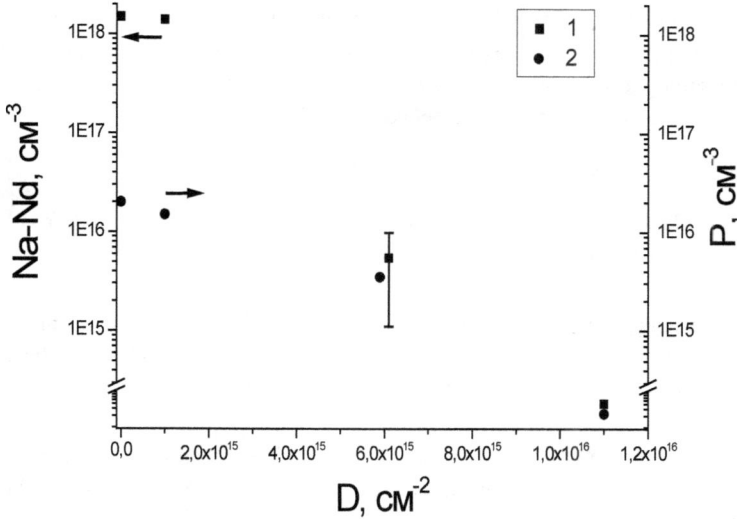

Fig. 2.2. Uncompensated acceptor concentration (1) and free carrier concentration (2) vs. the irradiation dose at room temperature [12].

The dependences in Fig. 2.2 can be used to estimate the carrier removal rate by using the known formula (2.2).

With expression (2.2) used, it is found that $\eta_e \approx 130$ cm^{-1} from the C - V data and $\eta_e \approx 1.8$ cm^{-1} on the basis of Hall data. This difference is due to the fact that these procedures measure different concentrations. In the case of capacitance-voltage characteristics, the concentration of uncompensated acceptors (atomic cores) present in the given layer is measured. From the Hall effect, we find the free carrier concentration at a given temperature. If the depth of acceptor (donor) levels is hundredths of eV (as, e.g., is the case for silicon), then Na - Nd will be equal to the free hole concentration at room temperature.

In WGSs of the SiC type, the ionization energies of the main impurity levels are substantially higher and these levels are not fully ionized even at room temperature, i.e., p < Na - Nd. A dependence of the ionization energy of the aluminum-related acceptor level on the concentration of aluminum is characteristic of SiC. This energy is 0.1 eV at Na - Nd $\approx 10^{19}$ cm^{-3} and increases to 0.24 - 0.27 eV at Nd - Na < 10^{19} cm^{-3}. Presumably, this is due to the formation of two different kinds of acceptor levels in SiC<Al> [15].

In this case, it is easy to estimate the would-be relationship between p and Na - Nd in our case. According to [16], in the case when Ea/kT >> 1 and the degree of compensation is small, the expression for the free carrier concentration has the form

$$P = ((Na-Nd)/2Nd)x \ Nv \ x \ exp(-Ea/kT) \tag{2.8}$$

Here, Ea = 0.24 eV is the ionization energy of the acceptor level, Nv = 2.5 10^{19} cm^{-3} is the density of states in the valence band, k is the Boltzmann constant, and T is absolute temperature.

In [12], the degree of compensation of our samples was considered to be small, ~10%. This was confirmed by the hole mobility of about 50 - 60 V/(cm^2 s) in the as-grown samples, which nearly coincides with the maximum hole mobility in 6H-SiC at this doping level [17].

Using expression (2.8) we obtain Na - Nd = 1.5 10^{18} cm^{-3} and p = 1.2 10^{16} cm^{-3}, which well coincides with the experiment. That is, p $\approx 10^{-2}$ (Na - Nd) for this conditions, or, with consideration for (2.1)

$$\eta_p = \{10^{-2} \ (Na-Nd)_0 - 10^{-2} \ (Na-Nd)_1\}/ D = 10^{-2} \ \eta_e \ (Na-Nd) \tag{2.9}$$

This means that the carrier removal rate found from C - V characteristics and Hall effect measurements will differ under the given conditions by two orders of magnitude.

2.4   Experimental data obtained in determining the value of η_e

The published data we are aware of, obtained in studies of the carrier removal rate in n- and p-type 4H and 6H SiC, are listed in Table 2.4. It can be seen that, in the case of electron irradiation of the n-type material, η_e is smaller for the sublimation-grown SiC, compared with the CVD SiC, and larger in the case of irradiation of a p-type material. At the same time, the carrier removal rates are approximately the same in materials with both conductivity types under irradiation with protons. Possibly, the difference in the rates is due to the secondary, rather than primary defect formation.

In our opinion, this difference may be due to the dissimilar spectra of DCs in SiC grown by different techniques. One of important distinctions of the CVD material from that grown by sublimation is in the small diffusion length of minority carriers in n-SiC.

It can be seen in Table 2.4 that this difference is by more than an order of magnitude. At the same time, there is no such difference in the p-type material, the diffusion length in p-type sublimation layers being occasionally even larger than that in CVD-grown layers. A possible reason for this difference is that the sublimation layers are grown at higher temperatures, when the growing layer is uncontrollably doped with aluminum and boron from the graphite crucible. These impurities form in SiC acceptor centers and compensate the n-type material. In addition, these centers are negatively charged in the n-type material and diminish the mobility and diffusion length of holes. At the same time, they are neutral in the p-type material and presumably have no effect on the mobility and diffusion length of electrons.

It is known that the electron irradiation of semiconductors results in that Frenkel pairs (vacancy-interstitial atom) are formed in these materials. It can be assumed that the presence of charged compensating acceptors in n-SiC grown by the SE method reduces the diffusion length of the components of a Frenkel pair and affects in the same way the diffusion length of carriers. This increases the recombination probability of the pair components, reduces the concentration of the radiation defects being formed, and improves the radiation hardness of the sublimation-grown n-SiC as compared with the material grown by the CVD method.

In irradiation of silicon carbide with protons having energies of 1 to 20 MeV, average energy of recoil atoms of silicon and carbon, (T_{av}) is 200 to 300 eV. Compared with the case of electron irradiation, there are two important distinctions. First, the higher T_{av} may lead to a substantial increase in the distances between the FP components being formed. Second, recoil atoms with above energies can cause collision cascades in which "secondary" FPs are formed, with the number of secondary FPs possibly being, according to TRIM, several times the number of primary FPs [19]. A cascade gives rise to microscopic regions 20 - 30 Å in size, in which up to ten displacements are generated. Strictly speaking, these microscopic regions are not yet disordered regions. However, the high concentration of vacancies in these regions favors formation of divacancies and complexes of these with impurity atoms.

If the hypothesis put forward in [12] is true, we can also prognosticate the behavior of the sublimation-grown SiC for other kinds of irradiation, with neutrons and γ-photons.

In the case of irradiation with 1 MeV neutrons, T_{av} is ~ 40 keV [18]. Recoil atoms with such energies generate along their range of ~500 Å [20] about a hundred displacements [16]. As a result, a set of point defects within a small local volume forms a single extended defect with specific properties, a disordered region (DR) [1]. Apparently, the presence of drains is unimportant for DRs and the behavior of sublimation- and CVD-grown SiC layers under neutron irradiation will be about the same.

Table 2.4. Carrier removal rates (η_e) and diffusion lengths of minority carriers in silicon carbide at room temperature in relation to the type of a material (polytype, conductivity type and fabrication technology) and to the kind of irradiation.

Kind of irradiation	Type of material							
	6H-SiC				4H-SiC			
	N		P		N		P	
	CVD	SE	CVD	SE	CVD	SE	CVD	SE
Electron energy 0.5 MeV							0,8 [25]	
Electron energy 1 MeV		0,015 [22]	0,26 [23]	1,3 [24]	0,1 [23]	0,015 [22]	1,6 [24]	1,7 [24]
Electron energy 6 MeV	1,0 [26]				2,15 [26]			
8 MeV protons	17 [11]	17 [11]			110-130 [11]	130 [11]		
450 keV protons					2×10^4 [27]			
1.7 MeV α-particles					4×10^5 [28-29]			
5.5 MeV α-particles	$7,8\ 10^4$ [30]							
Diffusion length of minority carriers in layers with concentration of ~ 10^{17} cm^{-3}, μm [17,32]	~ 10	~ 1	~ 1	~ 1,5	~ 12	< 0,5	~ 1,5	1-2

In the case of γ irradiation, e.g., in ^{60}C cobalt installations with average photon energy of 1.25 MeV, the main influence on the semiconductor is exerted by secondary Compton electrons with energy of ~ 0.7 MeV. It can be assumed that, under an irradiation of this kind, sublimation-grown layers will be more stable than those produced by CVD.

Thus, the appearance of microscopic regions with high concentration of vacancies in a single-crystal silicon carbide, in which reactions yielding divacancies and complexes of these with impurity atoms occur at a high rate, is the fundamental distinctive feature of the proton irradiation. The situation observed in the case of proton irradiation seems to be "intermediate" between the case of DRs under neutron irradiation and that of single point defects under electron irradiation. Specific features of sublimation-grown SiC layers will be strongly manifested under γ-photon and electron irradiation and hardly noticeable in the cases of proton and neutron irradiations.

It is noteworthy that, according to most reports, η_e (Si) $\geq \eta_e$ (SiC) at T = 300 K. The values obtained in [28, 29], for which the reverse is true, with $\eta_e(SiC)/ \eta_e(Si) \approx 10$, stand out from this body of data. In these studies, the irradiation was performed with 1.7 MeV α-particles at a dose of $2 \cdot 10^9$ cm^{-2}. The particle ranges were 3.8 μm in SiC and 5.9 μm in Si. It should be noted that the authors of these studies only monitored the effects occurring at the end of the particle range and, consequently, the effect may be due to helium formed at this depth, rather than to radiation defects.

Among the SiC polytypes, 6H and 4H SiC are the most promising for commercial application. Therefore studies concerned with the radiation hardness of other polytypes are not numerous. Some of the experimental data known to us are listed in Table 2.5. It can be seen from a comparison of Tables 2.4 and 2.5 that there is no significant dependence of η_e on the SiC polytype.

Table 2.5. Carrier removal rates (η_e) in other silicon carbide polytypes at room temperature

Kind of irradiation	Type of material	
	n-3C-SiC	p-15R-SiC
8 MeV protons	110 [33]	
1 MeV neutrons	7,2 [31]	1,0 [24]

In [33], the process of conductivity compensation in silicon and silicon carbide under irradiation with MeV electrons and protons was analyzed and the part played by recoil atoms in the formation of RDs was determined. For the CVD-grown n-4H-SiC silicon carbide, the formation of RDs leads to the conductivity compensation of the material. It was shown in [34] that the experimental values of the carrier removal rate (0.25 cm^{-1} at electron energy of 0.9 MeV) is nearly an order of magnitude smaller than that (2 cm^{-1}) for electrons with higher energy (6 - 8 MeV) [26, 35]. A similar relationship is also observed for silicon. The strong energy dependence of η_e cannot be due to the generation of primary RDs because the formation cross-section of primary RDs is nearly energy-independent within this range.

2.5   Compensation mechanism in SiC

The process of radiation defect formation in semiconductors includes two stages: generation of primary point defects of the vacancy - interstitial atom type (Frenkel pair, FP) and formation of secondary defects. The radiation defects (RDs) are generated via two processes. First, crystal-lattice atoms are displaced due to the direct interaction with an impinging particle, i.e., the so-called primary knocked-on atoms (PKAs) are formed. The second stage is named "formation of secondary RDs." In this stage, primary RDs interact with each other and with defects and impurities present in the semiconductor to give secondary RDs that are stable under the given conditions, and just these defects determine the change in the properties of the materials and devices.

The behavior of the Frenkel pair components formed via irradiation of the semiconductor depends on numerous factors, of which the following are the most important:

- impurity composition and doping level of a particular semiconductor;

- energy of a primary knocked-on recoil atom.

The many years' experience of experimental studies of the behavior of radiation defects upon their generation recommends creating conditions in which there is the main formation mechanism of RDs reliably recorded in experiments. These conditions are the following.

- Use of an irradiation type under which only genetically related intimate Frenkel pairs are generated. This is irradiation (frequently named "model irradiation") with electrons having the energy of the order of 1 MeV. In this case, PKAs have the energy of ~50 eV which is insufficient for the displacement cascades to be formed.

- Use of moderately doped materials (with a dopant concentration of ~10^{15} cm^{-3}).

So far, the compensation process has been simulated and the dose dependences of the conductivity compensation under irradiation with MeV electrons have been studied in detail only for a single elementary semiconductor, silicon [6]. The goal of our present study was to solve the above problems for the case of irradiation of a binary semiconductor, silicon carbide, with MeV electrons.

2.5.1 Model

Let us evaluate the kinetics of defect formation in the interaction of a single component of a Frenkel pair (in one of the silicon carbide sublattices) with impurity atoms of a single type. The concentrations of primary defects (e.g., vacancies in the silicon sublattice) (V) and the concentrations of free impurity atoms (N) vary with the irradiation duration (t) as follows:

$$dV/dt = \eta_{FP}G - V/\tau - \beta VN \qquad (2.10)$$

$$dN/dt = - \beta VN \qquad (2.11)$$

Here, V is the vacancy concentration; G, electron flux; η_{FP}, probability of Frenkel pair formation by a single electron, equal to the product of the defect-formation cross-section σ by the concentration N_{at} of regular silicon atoms

$$\eta_{FP} = \sigma N_{at} \qquad (2.12)$$

τ, vacancy lifetime determined by drains; and β, constant of vacancy interaction with an impurity atom. The first term in equation (2.10) describes the FP generation, the second term represents their recombination on insaturable drains, and the third term is responsible for their interaction with the dopant. Equation (2.11) contains a term describing the decrease in the concentration of free impurity atoms via their interaction with the FP components. It should be noted that a part (occasionally significant) of genetically related Frenkel pairs recombine immediately after being generated. Therefore, the fraction of vacancies (f_{FP}) capable of being involved in the formation of stable defects and, in particular, those with impurity atoms should be taken into account in expression (2.12). Commonly, this is limited to introducing as a parameter the effective formation cross-section of separated (dissociated) Frenkel pairs [36]:

$$\sigma_{eff} = \sigma f_{FP} \qquad (2.13)$$

In what follows, just this cross-section will be used in expression (2.12) when calculating the generation rate of separated Frenkel pairs (or the effective FP generation rate). The concentration N_c of complexes of secondary defects (vacancy - impurity atom) is calculated by the formula

$$N_c = N_0 - N \qquad (2.14)$$

31

This concentration is zero at the initial instant of time. Here, N_0 is the initial concentration of free impurity atoms. The carrier (electron) concentration n may vary with the irradiation dose in the case of deep trap formation due to the following two factors. The first of these is the capture of electrons by traps created by intrinsic FPs (e.g., by a vacancy). The second consists in that complexes (dopant atom--intrinsic point defect) are formed. If a complex being formed is neutral, then the equation for n takes the form:

$$n = N_0 - V - N_c = N - V \tag{2.15}$$

If the complexes are deep acceptors capturing electrons from the conduction band, we have, with consideration for (2.14), the following:

$$n = N - V - N_c = 2N - N_0 - V. \tag{2.16}$$

Formula (2.10) shows that, for an intrinsic defect (vacancy) to be predominant in the conductivity compensation mechanism, it is necessary to neglect the last two terms in expression (2.10). This can be done if the vacancy lifetime determined both by drains and by the impurity capture substantially exceeds the irradiation duration and is tens of minutes. In this case, the vacancy concentration linearly grows with the irradiation dose

$$V = \eta_{FP}D. \tag{2.17}$$

According to equation (2.11), the concentration of free impurity atoms remains unchanged and equal to N_0. At the same time, equation (2.15) shows that the carrier concentration decreases with increasing irradiation dose $(D = Gt)$:

$$n = N_0 - V = N_0 - \eta_{FP}D. \tag{2.18}$$

This equation describes the mechanism (we call it mechanism no. 1 or mechanism of compensation by intrinsic defects) according to which the carrier concentration decreases under irradiation due to the formation of singly charged deep acceptor centers in the semiconductor bulk, to which electrons go from the donor levels. In this case, the rate of electron removal from the conduction band (η_e) coincides with the effective generation rate of separated Frenkel pairs (η_{FP}). It is important to note that, in this model, the radiation defects forming deep centers may contain components of Frenkel pairs (vacancies and interstitial atoms) with only native semiconductor atoms. Thus, this mechanism is dominated by the contribution of primary RDs. Figure 2.3 shows separate straight lines which are solutions to equation (2.18) for the carrier concentration at various values of the parameter η_{FP}.

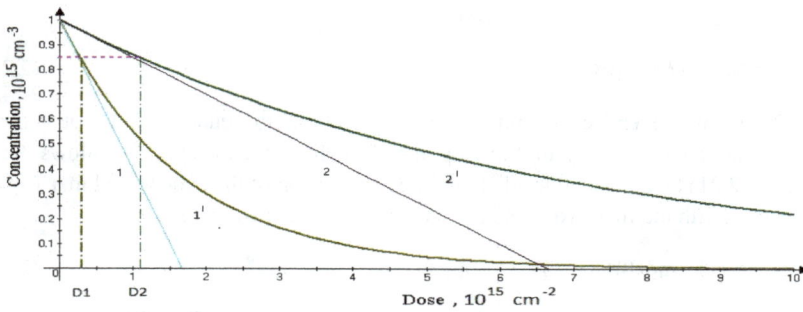

Fig. 2.3. Dependence of the carrier concentration on the electron irradiation dose.

Straight lines (1) and (2) represent a calculation by equation (9) at a parameter η_{FP} of (1) 0.6 and (2) 0.15 cm^{-1}, respectively. Curves (1') and (2') represent a calculation by equation (3.21) at a factor ($\eta_{FP} \beta \tau$) in the exponent equal to (1') 0.6 10^{-15} and (2') 0.15 10^{-15} cm^2, respectively [34].

The dashed line shows the level of "low" semiconductor compensation and the corresponding low-compensation doses for two samples, D_1 and D_2 [34].

In mechanism no. 2, the elementary defects formed in irradiation of the semiconductor effectively interact with dopant atoms to give neutral complexes (This mechanism is commonly regarded as "deactivation" or "passivation" of donor impurity states by radiation defects.) This occurs when the vacancy lifetime is substantially shorter than the irradiation time and is determined by drains. In this case, the vacancy concentration can be considered steady and can be found from equation (2.10) to be

$$V = \eta_{FP} G \tau. \tag{2.19}$$

In this case, the dependence of the carrier concentration on the irradiation dose is determined by the interaction of a vacancy with an impurity atom. Thus, the "dopant-passivation" mechanism is dominated by the contribution from secondary radiation defects, vacancy-impurity atom complexes. In this case, the carrier concentration is equal to the concentration of free impurity atoms, n = N.

Equation (2.11) can be used to calculate the variation kinetics of N. In the "passivation" mechanism, the kinetics is described by the equation

$$-dN/dt = \eta_{FP} \beta G \tau N. \tag{2.20}$$

33

The solution to this equation has the form

$$N = N_0 \exp(-\eta_{FP} \beta \tau G t). \tag{2.21}$$

Figure 2.3 shows several exponentials that are solutions to equation (2.21) at various values of the factor $(\eta_{FP}\beta\tau)$ multiplying the dose in the exponent. It follows from expression (2.21) that, at small irradiation doses, expansion of the exponential in a Taylor power series (with the first two terms of this expansion retained) gives

$$N = N_0 (1 - \eta_{FP} \beta \tau G t) = N_0 (1 - \eta_{FP} \beta \tau D), \tag{2.22}$$

i.e., the dose dependence of the compensation becomes linear. The differences of the exponentials from straight lines ("low" compensation level) start to be observed in Fig. 2.3 at $(n/n_0) \approx 0.85$. However, the dose at which the compensation in a particular semiconductor can be considered low is not a universal quantity, being dependent on the parameters η_{FP}, β, and τ of the radiation defects in a particular material. Strictly speaking, at small N_0, the depletion of "impurity drains" for vacancies and, consequently, departure of dependences (2.22) from linear behavior is observed even at very low irradiation doses (short durations).

Finally, the conductivity compensation mechanism no. 3 includes elements of the first two. The secondary defect complexes being formed are themselves deep acceptors effectively capturing electrons from the conduction band. The concentration of the free donor impurity can be still calculated by equations (2.21) and (2.22); however, the electron concentration already does not coincide with N, being rather calculated by the formula:

$$n = N - N_c = 2N - N_0. \tag{2.23}$$

Taking into account formula (2.21), we have

$$n = N_0 (2 \exp(-\eta_{FP} \beta \tau G t) - 1) \tag{2.24}$$

At low irradiation doses, formula (2.24) takes the form

$$n = N_0 (1 - 2 \eta_{FP} \beta \tau G t) = N_0 (1 - 2 \eta_{FP} \beta \tau D) \tag{2.25}$$

Comparison of the two formulas (2.22) and (2.25) shows that the slope ratio of the straight lines describing the concentration of conduction electrons is twice that for the straight lines describing the dopant concentration.

Let us specify that a vacancy formed under irradiation actually "removes" two, rather than one electron from the conduction band. Naturally, secondary radiation defects also play a principal role in this mechanism, too.

Our analysis shows that a study of the process of deep conductivity compensation under irradiation can, in principle, provide information about the mechanism of this phenomenon if the dependence of the effect on the irradiation dose (and intensity) is examined in detail.

Indeed, if the vacancy lifetime substantially exceeds the irradiation duration and the compensation is governed by vacancies, then the dose dependence of the compensation is linear to the point of full compensation.

If the vacancy lifetime is substantially shorter than the irradiation duration and the process is controlled by the interaction of a vacancy with an impurity atom, then the linearity of the dependence is disrupted at high doses in accordance with formulas (2.21) and (2.24).

Let us now pass to an experimental verification of the mechanisms under consideration for the case of SiC.

2.5.2 Comparison with experiment

Silicon carbide 4H-SiC samples having the form of 50-μm-thick epitaxial films grown by the CVD method in Germany (Leibniz-Institute for Crystal Growth, Berlin). The uncompensated donor (N_d- N_a) or acceptor (N_a - N_d) concentrations in the as-grown and irradiated samples were found from capacitance-voltage (C - U) characteristics measured at room temperature.

In [34], experiments were carried out with two kinds of 4H-SiC, of n- and p-types. Samples with approximately the same doping levels ($\sim10^{15}$ cm^{-3}) were chosen. All the samples were irradiated simultaneously.

The irradiation with 0.9 MeV electrons was performed on a resonant transformer accelerator (pulse repetition frequency 450 Hz, pulse width 330 μs) on a target cooled with flowing water. The range of 0.9 MeV electrons is ~1.0 mm in SiC. The average current density of the electron beam was 12.5 μA cm^{-2}. It can be assumed that defects were introduced under electron irradiation homogeneously throughout the sample volume because the thickness of the SiC samples being irradiated was substantially smaller than the range of electrons.

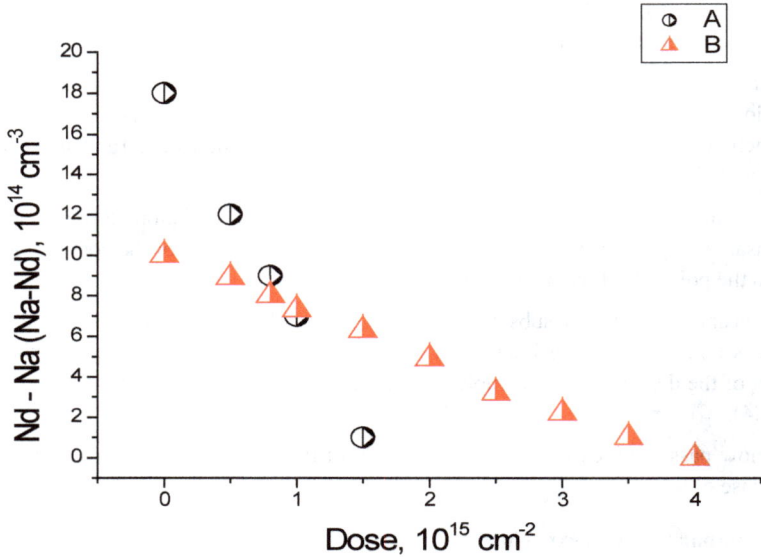

Fig. 2.4. Experimental dependences of the uncompensated carrier concentration on the electron irradiation dose: (B) N_a - N_d in p-4H-SiC, (A) N_d - N_a in n-4H-SiC [34].

Figure 2.4 shows the experimentally measured dose dependence of the electrical conductivity compensation in silicon carbide under electron irradiation: room-temperature concentrations (N_d- N_a) for n-CVD–4H SiC and (N_a - N_d) for p-CVD–4H SiC. It can be seen that a linear dose dependence is observed up to substantial semiconductor compensation levels (90%) for both p-SiC and n-SiC. This suggests that the carrier concentration decreases under irradiation due to the formation of deep centers in the semiconductor bulk, to which electrons from shallow donor levels or holes from shallow acceptor levels go. The radiation defects forming deep centers may contain components of Frenkel pairs (vacancies and interstitial atoms) with only native semiconductor atoms. The lifetime of isolated vacancies exceeds the irradiation time and constitutes tens of minutes at room temperature. It is noteworthy that the experimental dose dependences are independent of the irradiation intensity (at electron beam current densities of 1.25 to 12.5 μA cm^{-2}). The rate of electron removal from the conduction band (η_e) coincides in this case with the effective generation rate of separated Frenkel pairs

(η_{FP}). Figure 2.5 compares the calculated [by formula (2.18)] and experimental dose dependences for n-SiC. The best agreement is achieved at a formation rate of separated Frenkel pairs, $\eta_{FP} \approx 0.25$ cm^{-1}. For p-SiC, a coincidence is achieved at η_{FP} in the range of 1.2 - 1.6 cm^{-1}.

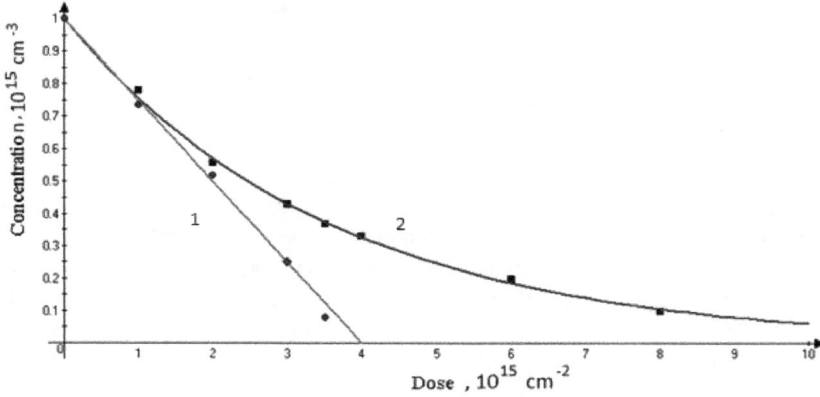

Fig. 2.5 Conductivity compensation in (1) n-SiC and (2) n-Si under irradiation with 0.9 MeV electrons. Points represent experimental data (SiC, present study; and Si, [36]). Straight line 1 represents a calculation by equation (3.18) at a parameter η_{FP} of 0.25 cm^{-1}. Curve 2 represents a calculation by equation (3.24) at a factor ($\eta_{FP} \beta \tau$) in the exponent equal to $1.2\ 10^{-16}$ cm^2 [34].

Let us evaluate the probability of separation of a Frenkel pair being formed in p- and n-type materials. The total FP generation rate η_{FP} is calculated as a product of the full cross-section σ of FP formation by the concentration of atoms in the target. Accordingly, the generation rate η_{FP} was obtained for SiC as a sum of partial values of η_{FP} for silicon and carbon atoms. Then, at a known value of $\sigma \approx 40$ barn [37], η_{FP} will be ~2 cm^{-1} for silicon in the sublattice of silicon carbide. For carbon at $\sigma = 23$ barn [37], η_{FP} in silicon carbide will be ~1 cm^{-1}. The total rate of FP formation in the SiC sublattices is ~3 cm^{-1}.

In SiC, the experimentally observed value of η_e (~0.25 cm^{-1}) is substantially smaller than the calculated η_{FP}(Si) and η_{FP}(C). Therefore, vacancies in both sublattices may be responsible for the conductivity compensation. It is believed in the literature that the main defects responsible for the compensation of n-SiC (CVD) are the Z_1/Z_2 center and EH$_{6/7}$ center [38]. To identify the sublattice in which these centers are formed, irradiation with 150 keV electrons is commonly performed. At these electron energies, defects are formed only in the carbon sublattice. It was demonstrated in [39, 40] that the above centers are reliably detected at bombarding-electron energies lower than 200 keV. Both centers are associated with the carbon vacancy [41]. If a carbon vacancy is operative, the separation probability of a Frenkel pair formed in this sublattice, f_{FP}, is ~25% according to formula (2.13). The remaining 75% of the FPs recombine immediately after being generated. Unfortunately, it is hardly possible to compensate n-SiC with ~150 keV electrons because this requires huge irradiation durations (due to the low defect introduction rate). In the studies known to us, the minimum electron energy at which the compensation was observed in n-SiC (CVD) is 400 keV [42].

In p-SiC, the experimentally observed value $\eta_p \approx (1.2 - 1.6)$ cm^{-1} (see Fig. 2.4) exceeds the calculated $\eta_{FP}(C)$. Therefore, either primary defects in the silicon sublattice or defects in both sublattices are responsible for the conductivity compensation. Unfortunately, in contrast to n-SiC, experimental studies concerned with irradiation of p-SiC are very scarce. If only the silicon vacancy is operative, the separation probability of a Frenkel pair in this sublattice, f_{FP}, is ~80% according to formula (2.13). The strong difference in FP behavior (degree of dissociation) between p- and n-type materials may be due not only to the different sublattices, but also to the charge state of the components of FPs formed in p- and n-SiC materials.

Figure 2.5 presents for comparison the experimental data for n-FZ-Si taken from [36]. These points exemplify the third mechanism, named Watkins mechanism, just which is operative in electron-irradiated silicon. In this case, the role of a deep acceptor, a negatively charged complex, is played by the E-center (containing a vacancy and a phosphorus atom). The experimental points are well described by curve (2.24) for the factor ($\eta_{FP} \beta \tau$) multiplying the dose in the exponent equal to 1.2 10^{-16} cm^2.

Let us estimate the vacancy lifetime for silicon. We take into account that the parameter β, the interaction constant of a vacancy with an impurity atom, is equal to the product of two parameters ($\beta = \alpha\gamma$). The parameter β depends on the velocity of vacancy motion over the crystal and on the interaction probability of a vacancy with an encountered impurity atom. The parameter α characterizes the motion velocity of the primary defect over the crystal, or, more precisely, the crystal volume covered by a vacancy in unit time. The value of α is proportional to the probability of encounter of a primary defect with a donor atom. The value of α can be estimated as $\alpha = a^3 v \exp(-E_m/kT)$, where a is the lattice constant; v, vibration frequency of atoms in the lattice; and E_m, vacancy migration energy. This estimate gives a value $\alpha = 10^{-9} \exp(-E_m/kT)$ for Group-IV atomic semiconductors.

Using the value $E_m = 0.18$ eV for silicon [43], we obtain for α a room-temperature value of $7.5 \ 10^{-13} \ cm^3 \ s^{-1}$.

The parameter γ is the probability of formation of a secondary defect upon encounter of a vacancy and an impurity atom. Taking into account the large cross-section of the interaction between a negatively charged vacancy and a positive phosphorus atom, we can take that γ is unity for this pair of interacting components.

The experimental points for the electron concentration (Fig. 2.5) are well described by the exponential (2.24), with the factor multiplying the dose in the exponent equal to 1.2 10^{-16}. Knowing the product of three factors ($\eta_{FP}\beta\tau$), the value of η_{FP} ($\sim 2 \ cm^{-1}$), and that of β ($7.5 \ 10^{-13} \ cm^3 \ s^{-1}$), we can estimate the vacancy lifetime to be $\tau \approx 80$ μs.

It should be noted that 4H-SiC (CVD) is presently the most perfect silicon carbide material. For comparison, we studied in [22] the rate of carrier removal, η_e, in n-SiC (SE) layers. We found that, in the case of electron irradiation, η_e is 0.015 cm^{-1} in sublimation-grown n-SiC and has an order of magnitude larger value of 0.25 cm^{-1} in CVD films. It is known that the generation rate of primary RDs is nearly independent of the method used to grow the material [44]. Therefore, it is necessary to assume that the difference in the degradation rates of the material is accounted for not so much by the primary defect formation, as by that of the secondary type. The specific features of the fabrication technique created in the sublimation material favorable conditions for the drain and recombination of point defects, which leads to low carrier removal rates under electron irradiation.

2.6   Radiation doping

Another important aspect of studies concerned with the interaction of ionizing radiation with substance is the radiation doping (RDo). The RDo of a semiconductor can create local regions with high resistivity, which are necessary for developing a device. It is commonly believed that the radiation hardness and the possibility of RDo are contradicting characteristics of a material, with a radiation-hard semiconductor considered to be unpromising for compensation by RDo.

We show below that such an assessment is not quite correct in the case of WBSs. It is known that WBSs are promising not only for high-temperature electronics, but also for a number of devices not intended to operate at high temperatures. In the first place, these are high-frequency devices (Schottky diodes, some kinds of FETs) the structure of which contains a metal - semiconductor contact. As a rule, the contact of this kind rapidly degrades with increasing temperature, irrespective of the maximum possible working temperatures of the semiconductor itself. If the working temperatures of a device do not

$$n = \frac{N_D N_C}{N_{RD} - N_D} \exp\left(-\frac{E_{RD}}{kT}\right) \qquad (2.25)$$

exceed substantially 300 K, then we can use the values η_e obtained at this temperature for determining the RDo efficiency. Another important parameter characterizing the RDo efficiency is the maximum resistivity (ρ_M) that can be obtained for various semiconductors. This parameter can be evaluated by the known formula

$$\rho_M = (e n \mu)^{-1} \qquad (2.26)$$

Here, e is elementary charge, n is the concentration of electrons in the conduction band, and μ is the carrier mobility.

Let us assume, when carrying out a calculation, that an acceptor RD was formed at the midgap ($E_{RD} = 1/2 E_g$) upon irradiation of an n-type material, with its concentration N_{RD} exceeding the total concentration of shallow donors. Then n can be evaluated by the following formula [45]:

$$n = N_D N_C (N_{RD} - N_D)^{-1} \exp(-E_R D/kT) \qquad (2.27)$$

The results obtained in calculating the ratio between the resistivity of some semiconductors and that of silicon at T = 300 K are presented in Table 2.6. It can be seen that WGSs markedly surpass NGSs as materials for obtaining high-resistivity layers, in agreement with the conclusions of [46]. The semi-insulating properties of the layers will also be preserved at elevated temperatures. It was shown in [10] that the resistivity of 4H SiC at T = 450 K is equal to the maximum value obtained for GaAs at room temperature ($\rho \approx 10^9$ Ω cm).

Table 2.6. Calculated ratio between the maximum resistivity of wide-bandgap semiconductors and that of silicon at 300 K

Semiconductor	ρ/ρ_{Si}
GaAs	2.4×10^4
6H-SiC	6.6×10^{15}
4H-SiC	3.5×10^{18}

In paper [47] electron irradiation was applied to the formation of 4H-SiC semi-insulating layers. The resistivity of irradiated layers (energy 400 keV, fluence ~ 1,9 x 10^{18} cm^{-2}) exceeded 10^{10} Ω cm at room temperature. From capacitance-voltage characteristics of Schottky structures, authors estimate the depth of semi-insulated layers as 10 μm. The semi-insulating property can be ascribed to electron trapping at the $Z_{1/2}$ and $EH_{5/7}$ centers generated by electron irradiation.

2.7   Effect of high irradiation doses

The amorphization of SiC samples has been observed upon irradiation with electrons [4, 48], neutrons [49], protons, and various kinds of ions (He, Ar, Cr) [50]. It was also noted in [50] that the process of SiC amorphization differs from that for silicon. Only ions heavier than B can amorphize silicon at room temperature, with lighter ions yielding strongly defective layers of the crystalline material even at very high implantation doses. At the same time, the amorphization of SiC occurs when the critical implantation density of 23 eV/atom is reached, irrespective of the mass of an ion. It was noted in [50] that this difference may be due to the higher room-temperature mobility of defects in silicon and their partial annealing (recombination) directly in the track of an implanted ion.

It is well known that the depth distribution of stopping energy losses for protons in a material is quite specific. As a result, radiation defects are produced by protons in a semiconductor predominantly at depths that are close to the protons' projected range Rp. It is important that the radiation-defect generation rate directly at the surface is almost an order of magnitude lower than that at depths close to Rp. At the same time, radiation defects are not produced at all at depths that far exceed Rp [44]. Thus, irradiation with protons can be used to form layers that are highly localized at a certain depth (buried layers) and have a high concentration of radiation defects. As a result, irradiation with high doses of low-energy protons is widely used in the technology of various semiconductors, for example, to form highresistivity buried layers (SOI technology) or to detach thin films of material after annealing of irradiated samples (so-called Smart Cut technology) [44]. The use of irradiation with protons in SiC technology may be of particular interest in the context of a recent report that it is possible to form an amorphous SiC layer as a result of exposure to proton radiation [51]. As is well known, the phenomenon of amorphization under the effect of irradiation with protons is typically not observed for other semiconductors [44].

In [52] was studied the formation of buried SiC layers with a high concentration of radiation defects under low energy proton irradiation. Used were $6H$- and $4H$-SiC wafers with n-type conductivity as the samples. The concentration of uncompensated donors $Nd - Na$ was ~1.6 x 10^{18} cm^{-3} in $6H$-SiC and 3.8 x 10^{18} cm^{-3} in $4H$-SiC. The samples were irradiated with protons using the accelerator of an NG-200U neutron generator at room temperature. We used protons with energy $E = 100$ keV and with doses ranging from 2 x 10^{17} to 4 x 10^{17} cm^{-2}. The value of Rp for 100-keV protons is equal to ~0.65 μm in SiC [44]. The samples were subjected to post irradiation isochronous annealing at temperatures in the range 500– 1500 °C in an argon atmosphere. The parameters of the samples before and after irradiation and after annealing at various temperatures were determined from measurements of the capacitance–voltage (C–V) and current–voltage (I–V) characteristics, as well as from an analysis of photoluminescence (PL) spectra.

A band peaked at a wavelength of $\lambda_{max} \sim$ 465– 470 nm was observed in the PL spectra of unirradiated 6H-SiC samples. This band is caused by radiative recombination at the Al–N donor–acceptor pairs.

A PL band peaked at $\lambda_{max} \sim$ 420 nm was also observed in the PL spectrum of unirradiated 4H-SiC crystals; this band is related to radiative recombination involving the Al acceptor center. In addition, a band peaked at $\lambda_{max} \sim$ 560 nm and related to the B impurity in SiC crystals [53,54] was observed in the long-wavelength portion of the 4H-SiC PL spectrum (see section 3.5.1 for more details). The PL almost disappeared in all types of samples immediately after irradiation. Previously [55-57], the quenching of PL related to the atoms of the Al impurity was studied in SiC samples treated in a hydrogen plasma. This quenching was presumably [55-57] caused by the ability of hydrogen atoms to passivate the B and Al acceptor centers in SiC, i.e., the ability of hydrogen to form electrically inactive complexes with the Al and B atoms. These complexes do not exhibit a high thermal stability and decompose at temperatures >200 °C; simultaneously, diffusion of hydrogen from the semiconductor bulk is observed [57]. In [55] was observed the complete recovery of the initial PL intensity after annealing samples at temperatures ~750 °C. In case [52], an appreciable recovery of the PL intensity sets in at higher temperatures (~800 °C), which is apparently related to the much higher hydrogen concentration in our samples than in the samples studied in [57]. It is worth noting that this temperature (~800 °C) is close to the temperature for the onset of annealing of radiation defects in SiC [58, 59, 11]. The PL intensity in the samples under study reverted to the initial (preirradiation) value after annealing at 1500 °C. Studies of the surface of irradiated samples showed that morphological parameters remained unchanged after irradiation with protons in the entire range of doses used. A variation in the surface morphology sets in after annealing of irradiated samples. It is worth noting that, in contrast to silicon [51], the annealing temperatures required are quite high (~880 °C).

Blistering and detachment of flakes of the material were observed for all the samples irradiated with a dose $< 3 \times 10^{17}$ cm^{-2} of protons and then annealed at temperatures ≥ 800 °C [2]. Blistering of the surface was not observed for the samples irradiated with a dose $\geq 3 \times 10^{17}$ cm^{-2} of protons and then annealed in the temperature range 800–1500 °C. In our opinion, the observed suppression of blistering is caused by amorphization of the implanted layer. In [51] was used atomic-force microscopy to detect the presence of an amorphous phase in silicon carbide irradiated with low energy protons with a dose of 3

$\times 10^{17}$ cm^{-2} and comparatively low proton-beam currents (<5 µA cm^{-2}). Apparently, this observation is indicative of specific features of separation of primary radiation defects in silicon carbide, which may lead to the accumulation of a concentration of multivacancy complexes that is sufficient for amorphization of the semiconductor. Amorphization suppresses the formation of flat two-dimensional vacancy–hydrogen defects that are observed only in the crystalline material and represent the main factor in the development of microcracks in the plane of buried defect-rich layers, which gives rise to blistering.

A shift of the $C–V$ characteristics upward along the vertical axis with the initial slope (the initial value of $Nd – Na$) retained was observed after irradiation of the surface-barrier structures under study. In addition, a dependence of the measured capacitance C on the frequency used in the measurements (f) was observed: C increased as f decreased. The value of C coincided with that for an unirradiated diode at frequencies ~100 Hz. This feature of the $C–V$ characteristic [60-61] indicates that the ohmic resistance of the diode increases significantly a result of irradiation. In addition, a nonlinear dependence $C^{-2}(U)$ was observed for 6H-SiC samples at reverse-bias voltages > 8 V (thickness of the space charge layer >0.1 µm); this observation indicated that the difference concentration $Nd – Na$ decreased with increasing depth. This effect is much less pronounced for the 4H-SiC samples, which is evidently related to the higher initial (before irradiation) concentration $Nd – Na$ in the samples of this polytype (thickness of the space charge layer is ~0.06 µm at the highest reverse-bias voltages). Thus, according to the capacitance measurements, irradiation led to the formation of a compensated layer at depths >0.1 µm; at the same time, the concentration at the surface of the sample did not change.

Finally in [52] was concluded that dose of ~3 × 10^{17} cm^{-2} is the critical dose for 100-keV protons. Irradiation with doses <3 × 10^{17} cm^{-2} leads to the development of blistering during annealing and can be used in Smart Cut technology. This blistering phenomenon is found to be suppressed at doses >3 × 10^{17} cm^{-2}, most likely because of the developing amorphization of the irradiated layer. The thickness of this layer can be varied by varying the proton energy. Thus, irradiation with protons with doses > 3 × 10^{17} cm^{-2} may turn out to be promising for the formation of SOI structures based on SiC. It is worth noting that the cross section for the defect production usually decreases as the proton energy increases; as a result, the concentration of defects in the buried layer decreases. Therefore, we may expect that an increase in the proton energy will lead to an increase in the critical value of the radiation dose.

2.8   Effect of the energy of recoil atoms on conductivity compensation in moderately doped n-Si and n-SiC under irradiation with MeV electrons and protons

2.8.1   Introduction

The irradiation of semiconductors with MeV electrons and protons is used both for basic research purposes and for the solution of applied problems. Scientifically, it is rather important to verify the existing opinion that the interaction of intrinsic primary radiation defects in semiconductors under irradiation with fast electrons occurs similarly to the case of irradiation with MeV protons. The main practical application field is the technique for reducing the lifetime of nonequilibrium carriers in structures of power semiconductor (as a rule, based on silicon and silicon carbide) devices of the switching type: diodes, transistors, and thyristors [62]. It is impossible not to note the importance of studies carried out with beams of MeV electrons and protons for simulating the space conditions [63] and, in particular, the Earth's radiation belts [64].

It is known that the radiation-induced defect formation is accompanied by the appearance of local energy levels in the band gap of a semiconductor. Radiation defects are for the most part either compensation centers (carrier "traps"), or recombination centers for nonequilibrium carriers [34]. Therefore, irradiation can widely vary the characteristics of semiconductors (concentration, mobility, and lifetime of carriers) and the properties of semiconductor devices [44]. As is known, radiation defects (RDs) are formed in semiconductors by bombarding particles in two stages. The first of these, "generation of primary RDs," in turn, includes two processes. First, host atoms are shifted as a result of a direct interaction with a particle, with the so-called primary knocked-on atoms (PKAs) formed. Further, there occurs a cascade of collisions in which defects are created by PKAs themselves. In the second stage, "formation of secondary RDs," the primary vacancies and interstitial atoms enter into various physicochemical reactions (mainly with the defects and impurities present in the matrix) to give stable RDs, and just these RDs further determine changes in properties of materials and devices.

For silicon carbide n-4H-SiC grown by vapor-phase epitaxy, formation of RDs leads to a conductivity compensation of the material. It was shown in [34] that experimental values of the carrier removal rate η_e (0.25 cm^{-1} at electron energy of 0.9 MeV) are nearly an order of magnitude lower than those (2 cm^{-1}) for higher energy electrons (6 - 8 MeV) [35,65 - 66].

It is noteworthy that the same capacitance-voltage method was used in [34, 35, 65 - 66] to determine the concentration. For another semiconductor, silicon, the dependence of the conductivity compensation on the energy of bombarding electrons was studied in [67,

68]. It should be noted that the same Hall method served in both studies to measure the carrier concentration. It was shown that η_e became an order of magnitude higher as the electron energy increased from 2 to 7 MeV. The strong energy dependence of η_e cannot be attributed to the process in which primary RDs are generated because the formation cross-section of primary radiation defects is nearly energy-independent in this energy range. So far, due attention has not been given to explaining the differences between the above experimental and calculated data.

Below will be done analyze of the process of conductivity compensation in silicon and silicon carbide under irradiation with MeV electrons and protons and to determine the role played by recoil atoms in the formation of radiation defects.

Let us consider the process in which radiation defects are formed, yielding compensation centers in semiconductors under these kinds of irradiation.

2.8.2 Generation of primary radiation defects under electron irradiation

The scattering in collisions leading to displacements is primarily associated with the Coulomb interaction of an electron with a host nucleus in the target, but Rutherford's nonrelativistic scattering laws are inapplicable in the given case because an electron with energy $E_e = 0.9$ MeV has a speed close to that of light ($V = 0.94c$). The relativistic Coulomb scattering of electrons shows that, with increasing energy of bombarding electrons, the displacement cross-section gradually grows from zero at a certain energy E_0 of electrons (in irradiation of silicon, the energy $E_0 = 250$ keV) and then becomes constant. Because the values of E_e analyzed in this study substantially exceed E_0, the scattering cross-section can be evaluated by the simplified McKinley - Feshbach formula [69]:

$$\sigma_d = (\tfrac{1}{4}\pi\varepsilon_0)^2 \left(\frac{2\pi Z^2 e^4}{E_d Mc^2}\right) = \left[\frac{140 Z^2}{A E_d (eV)}\right] \text{(barn)} \qquad (2.28)$$

where c is the speed of light; e, elementary charge; M and Z, mass and charge of a host atom in the target; A, mass number of this atom; and E_d, threshold energy of atomic displacement.

For Si, E_{th} was taken to be 25 eV, which is the value most frequently used in analyses and calculations in the literature [70]. For SiC, comparatively new values of 24 eV and 18 eV were taken for the silicon and carbon sublattices, respectively [71, 40]. At these threshold energies, the cross-sections of RD formation, given by formula (2.28), are ~ 40 barn for Si and 23 barn for C, being nearly independent of the energy of bombarding electrons in the range 1-10 MeV. The rate of PKA generation due to the interaction with an incident electron is calculated as a product of the cross-section by the concentration of host atoms in the target. Accordingly, the PKA generation rate for the binary semiconductor SiC was obtained as a sum of the partial generation rates for silicon and carbon atoms. Then, the PKA generation rate will be ~2 cm^{-1} for silicon and ~1 cm^{-1} for carbon. The summarized PKA generation rate in the SiC sublattices (3 cm^{-1}) is nearly the same within the range from 1 to 10 MeV. Thus, the rates of generation of primary radiation defects due to the interaction with an incident electron are approximately the same at energies of 1 and 10 MeV and cannot account for the observed difference between the carrier removal rates η_e. The reason for this difference should be sought for in the defect generation by recoil atoms, rather than by the bombarding electrons, and also in the formation of secondary radiation defects.

2.8.3 Formation of secondary radiation defects

The behavior of the components of a Frenkel pair (vacancy and interstitial atom), formed upon irradiation of a semiconductor, depends on numerous factors, the most important of which are the irradiation temperature, impurity composition and doping level of a particular semiconductor, and energy of a primary knocked-on recoil atom.

The many-years' experience of experimental studies of the behavior of radiation defects upon their generation recommends creating conditions in which only a single parameter is varied. In our studies [34, 68], only the last of these parameters was variable, i.e., the energy spectrum of PKAs, in contrast to [35, 65 - 67].

It is known that the energy spectrum of PKAs, or the number of atoms primarily knocked-on by a relativistic electron from their equilibrium positions, is distributed, as also in the case of bombardment with atomic particles, approximately by the inverse-square energy law [72]. Let us estimate the average energy $<E_{PKA}>$ received by PKAs for the example of a silicon atom colliding with a relativistic electron. Taking into account the inverse quadratic dependence for calculating the average energy received by semiconductor atoms in collisions with relativistic electrons, we can use the formula derived for the elastic Rutherford scattering [72]:

$$< E_{PKA} > = \left[E_d \frac{E_{max}}{E_{max} - E_d} \right] \ln \left(\frac{E_{max}}{E_d} \right) \tag{2.29}$$

where E_{max} is the maximum energy transferred to a semiconductor atom by a relativistic electron with mass m_e and energy E_e:

$$E_{max} = 2E_e(E_e + 2m_e c^2)/M c^2 = \frac{E_e(E_e + 1.022)}{469A} (MeV) \tag{2.30}$$

Table 2.7 list the average and maximum energies of silicon recoil atoms in relation to the energy of bombarding electrons. For comparison, similar parameters are given in Table

2.8 for the case of proton irradiation. In this case, calculations were made in two ways. Either analytically, by expression (2.29) in which E_{max} is estimated by the formula [15] for proton with mass m_p and energy E_p:

$$E_{max} = \frac{4m_p M E_p}{(m_p+M)^2} = \frac{4AE_p}{(1+A)^2}$$ (2.31)

or by numerical simulation of the deceleration of protons by TRIM software [16].

Table 2.7. Dependence of the main energy parameters of silicon recoil atoms and generation rate of Frenkel pairs in relation to the energy E_e of bombarding electrons.

E_e,	E_{max}	$\langle E_{PKA} \rangle$	ν	η_{FP},
MeV	eV			cm^{-1}
0.9	130	51	1.0	2.0
2.0	460	76	1.2	2.4
2.5	666	86	1.4	2.8
4.0	1530	105	1.7	3.4
8.0	5496	135	2.2	4.4
15.0	18500	160	2.7	5.4

Note. The following values were taken for Si: $E_d = 25$ eV, $N_0 = 5 \cdot 10^{22}$ cm^{-3}.

Table 2.8. Dependence of the main energy parameters of silicon recoil atoms and characteristics of Frenkel pairs on the energy of bombarding protons.

E_p, MeV	0.5	1	2	4	8	15	20
E_{max}, keV	67	134	368	535	1070	2006	2680
$\langle E_{PKA} \rangle$, eV	197	215	232	250	267	283	290
ν	3.1	3.4	3.7	4.0	4.3	4.5	4.6
σ_d, barn	37000	18500	9200	4600	2300	1200	900
η_{FP}, cm^{-1}	5700	3100	1700	920	494	280	207

Figure 2.6 shows how $<E_{PKA}>$ depends on the energy of bombarding electrons and protons. It is noteworthy that the rise in the average energy is more pronounced at E_e in the range from 1 to 4 MeV, compared with that at $E_e > 5$ MeV. Comparison of the curves shows that at higher electron energies (15 MeV), $<E_{PKA}>$ of 160 eV is considerably closer to the values obtained in the case of a proton irradiation (215 eV at energy of 1 MeV) as compared with those for the electron irradiation (55 eV at energy of 1 MeV). Therefore, it can be assumed that the RD spectrum will vary with increasing energy of bombarding electrons, thereby approaching the spectrum of RDs formed under proton irradiation.

It can be seen in Table 2.7 that, in the case of irradiation with 0.9 MeV electrons, $<E_{PKA}>$ of silicon is ~51eV, which is comparable with energy threshold E_d of defect formation. Thus, the electron energy of 0.9 MeV is sufficient for only single closely spaced (occasionally named "genetically related") pairs to be formed, constituted by a vacancy and an interstitial atom (Frenkel air, FP). In radiation physics of solids, this kind of irradiation is frequently named model irradiation. With increasing energy of bombarding electrons, a primary knocked-on atom can itself knock-on a lattice atom: the average number ν of displaced atoms per each PKA grows. Tables 2.7 and 2.8 present values of ν (also named multiplication coefficients), evaluated by the Kinchin-Pease formula [74] modified by Ziegler [73]:

$$\nu = \frac{<E_{PKA}>}{2,5E_d}, \qquad at \quad 2,5 << E_{PKA} > < E_i \qquad (3.32)$$

where E_i is the threshold ionization energy of an atom (equal to 7.15 keV for Si). The threshold energy level of a collision cascade, 62.5 eV for silicon, is shown by the dashed line in Fig. 2.6.

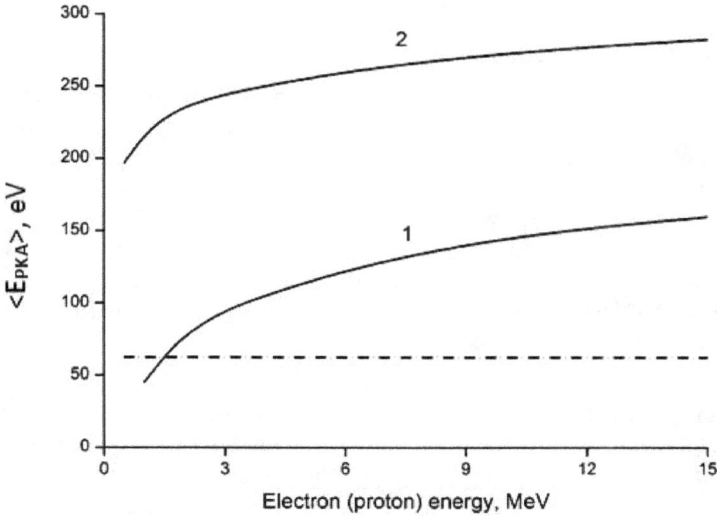

Fig. 3.6. Average energy of a primary knock-on silicon atom (PKA) on the energy of a bombarding particle. Curve 1: electron irradiation, calculation by formula (2.30). Curve 2: Proton irradiation, calculation by TRIM software [73]. The dashed line shows the onset energy level of a collision cascade (<E_{PKA}> = 2.5 eV; E_d = 62.5 eV) [84].

Formulas (2.28) and (2.29) show that, with increasing electron energy, the cross-section of defect formation remains unchanged, but <E_{PKA}> grows and, consequently, so does v. For example, for 8 MeV electrons, <E_{PKA}> = 135 eV and v ≈ 2.2. The concentration of displaced atoms (Frenkel pairs) formed due to the interaction with bombarding electrons and with PKAs under irradiation at a dose D is calculated by the formula

$$N_{FP} = N_{Si}\, \sigma_d\, v\, D \qquad (2.33)$$

where N_{Si} is the concentration of silicon atoms. It is more convenient to use the quantity named the FP generation rate η_{FP} introduced as the ratio of N_{FP} to the irradiation dose D. Then

$$\eta_{FP} = N_{Si}\, \sigma_d v \qquad (2.34)$$

49

Because, as calculated above for 0.9 MeV electrons, $\sigma_d = 40$ barn and $\nu = 1$, the PF generation rate found from (2.34) will be 2 cm^{-1} (Table 2.7).

2.8.4 Comparison with experiment

Silicon samples were cut from n-Si wafers. n-Si grown by the floating zone technique was doped with phosphorus in concentrations of $(6 - 8) \cdot 10^{15}$ cm^{-3}. The residual concentration of oxygen did not exceed $5 \cdot 10^{16}$ cm^{-3}. n-SiC (4H) epitaxial layers with thickness of 50 μm were grown in a commercial horizontal hot-wall CVD-system at the Leibniz Institute for Crystal Growth, Berlin, Germany. Commercial SiC (4H) wafers were used as substrates. The electron concentration due to uncompensated donors in these layers, $n = N_D - N_A$, did not exceed $2 \cdot 10^{15}$ cm^{-3}. Schottky diode structures were fabricated on the SiC layers. The irradiation and sample measurement procedures used in the present study were described in detail in [34, 68, 75].

Figure 2.7 shows experimental carrier removal rates η_e in silicon and silicon carbide, obtained in the present study for identical samples for two electron energies of 0.9 and 3.5 MeV, and those from [35, 65 - 67] for the energy range 2-8 MeV (curves 1 and 2). It can be seen that the values obtained for η_e under irradiation with 8 MeV electrons are approximately an order of magnitude higher than those for 0.9 MeV electrons. The same figure presents dependences of the calculated rate of Frenkel pair formation (η_{FP}) for silicon and silicon carbide (sum for both sublattices) on the energy of bombarding electrons (curves 3 and 4). It can be seen that the calculated rate of Frenkel pair formation under irradiation with 8 MeV electrons is only twice that for 0.9 MeV electrons.

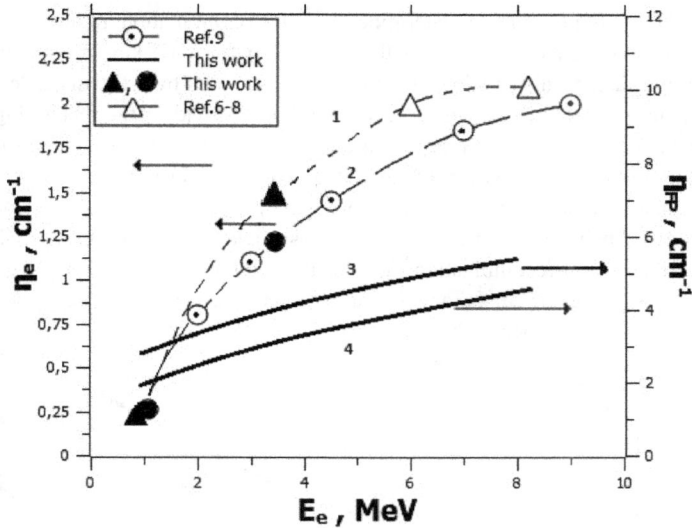

Fig. 2.7. Experimental values of the carrier removal rate in (1) silicon carbide and (2) silicon, obtained in the present study and in [6-9]. Dependence of the generation rate η_{FP} of Frenkel pairs for (3) silicon carbide and (4) silicon; calculation by formula (2.33) [84].

Thus, it is rather difficult to account for the experimental data by a simple increase in the generation rate of radiation defects due to the interaction with incident particles and PKAs.

To explain the experimental data, it is necessary to consider two more reasons. As the energy of bombarding electrons increases from 1 to 6 - 8 MeV, only the portion corresponding to head-on collisions changes in the PKA spectrum. This portion of "high-energy" ("hot") PKAs extends approximately from 150 to 1000 eV. With increasing PKA energy, the average distance between the genetically related FP components grows. Figure 2.9 shows how the range of PKA silicon ions in silicon depends on the PKA energy in the interval from 50 to 1200 eV, calculated by TRIM software [73]. It is known that the efficiency of defect formation is primarily determined by the extent of dissociation (separation) of primarily created genetically related Frenkel pairs: vacancy (V) - interstitial atom (I). In the course of dissociation, the charge of components of the pair may change. This circumstance determines the nature of the interaction between V and I and makes it possible to determine the FP recombination radius r_a (distance between the FP components, critical for their recombination). Of key importance in this context are two factors: distribution of primarily created FPs over the distance between the components and the presence of electrons and holes providing the recharging. It was noted in [76, 77] that V and I are neutral at the instant of birth, and the $[V^0 I^0]$ recombination radius is ~2a, where a is the lattice constant (for silicon, $a_{Si} \approx 0.543$ nm). When the vacancy captures an electron, the pair $[V^- I^0]$ dissociates at $r_a > 4a$. Upon a full recharging of the pair, $[V^- I^+]$, the distance increases and $r_a > 7a$. Figure 2.8 shows by dashed lines the critical recombination radii for these three cases. Silicon PKA energies corresponding to ranges of 2a, 4a, and 7a can be found from Fig. 2.8 These energies are 110, 420, and 1150 eV, respectively.

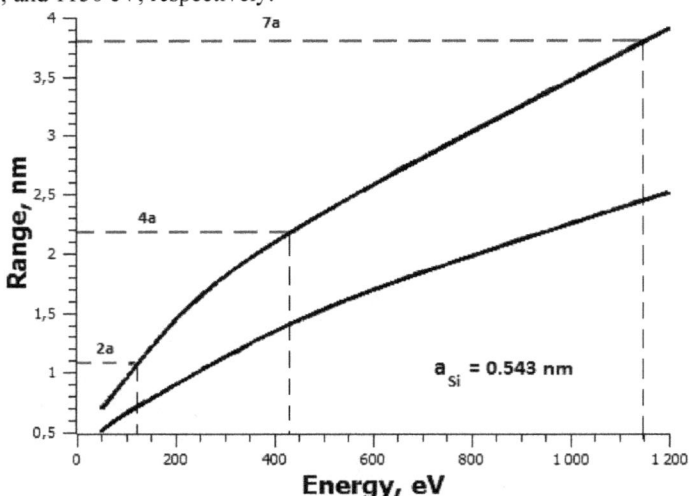

Fig. 2.8 Range of silicon PKAs in silicon (upper curve) and silicon carbide (lower curve) on the PKA energy, calculated by TRIM software [73]. Ashed lines indicate the values of the recombination radius and the corresponding energies of PKAs (see text)[84].

Thus, most of "hot" PKAs will be removed from their formation places and the Frenkel pair will separate into a free vacancy and an interstitial atom. As a consequence, the fraction f_{FP} of dissociating FPs grows with increasing energy of recoil atoms:

$$f_{FP} = \frac{\eta_e}{\eta_{FP}} \qquad (2.35)$$

The carrier removal rates in n-Si and n-SiC can serve in the initial irradiation stages as a measure of formation in silicon and silicon carbide of these "remote" pairs, i.e., those separated into isolated vacancies and interstitial atoms. It was known from the literature that, for example, the fraction f_{FP} of unrecombined FPs is ~5% in the case of irradiation of moderately doped silicon ($n \approx 10^{16}$ cm^{-3}) with 1 MeV electrons [77]. The experimentally measured rate of formation of dissociated pairs in silicon carbide (0.25 cm^{-1}) gives close values: f_{FP} is 8% of the total FP generation rate of 3 cm^{-1}.

As regards the identification of the radiation defects responsible for the conductivity compensation in moderately doped n-Si and n-SiC, the following comment should be made. In n-Si, the vacancy is believed to be the main component of a Frenkel pair, responsible for the compensation [67, 68, 78]. In silicon carbide, the situation is considerably more complicated. It is believed in the literature that the main defects responsible for the compensation in n-SiC (CVD) are the Z_1/Z_2 center and EH$_{6/7}$ center [38]. To identify the sublattice in which these centers are formed, n-SiC is commonly irradiated with 150 keV electrons. At these electron energies, defects are formed only in the carbon sublattice. It was shown in [39, 40] that the above-mentioned centers are reliably recorded at bombarding electron energies lower than 200 keV. Both the centers are associated with the carbon vacancy [41]. If only the carbon vacancy is responsible for the conductivity compensation, then the separation probability f_{FP} of the Frenkel pair formed in this sublattice is ~25% according to formula (2.35). The remaining 75% of the FPs formed recombine immediately after being generated. Unfortunately, it is hardly possible to perform compensation of n-SiC with electrons having energies of about ~150 keV because this requires huge irradiation times due to the low defect introduction rate. In the publications we are aware of, the minimum electron energy at which the compensation of n-SiC (CVD) was observed is 400 keV [42].

Let us estimate for the case of silicon irradiation with 0.9 MeV electrons the characteristic energy of PKAs in silicon (E_{char}) at which 5% of the total number of PKAs are in the right-hand ("high-energy") part of the spectrum. We take into account that the energy spectrum of PKAs, or the number of atoms primarily knocked-on by relativistic electrons from their equilibrium positions, is approximately distributed by the inverse-square energy law. With this dependence, it is easy to estimate the fraction of PKAs in the right-hand part of the spectrum (fraction of separated FPs):

$$f_{FP} = \frac{\frac{1}{E_{char}} - \frac{1}{E_2}}{\frac{1}{E_1} - \frac{1}{E_2}} \qquad (2.36)$$

where E_1 = 25 eV (threshold energy of defect formation), and E_2 = 130 eV is the maximum PKA energy estimated by formula (2.30). Under these conditions, the characteristic energy E_{char} is 107.5 eV. As already noted, the range equal to two lattice constants corresponds to this energy. This value of the annihilation radius may point to the annihilation of the genetically related components of Frenkel pairs in the neutral state $[V^0 \, I^0]$. As the energy of bombarding electrons increases, the fraction of PKAs that received energy higher than 107.5 eV becomes larger. For example, this fraction is about 20% according to formula (2.36) at the energy of 3.5 MeV. In this case, the experimentally measured ratio between the carrier removal rate and the rate of Frenkel pair generation is about 40%.

As the electron energy is raised to 6 MeV, f_{FP} in silicon carbide becomes nearly four times larger (increases to 30%) [35, 66]. It is noteworthy that, as the electron energy is raised further to 8 MeV, no significant rise in f_{FP} is observed. These results are correlated with the known data on the dependence of the conductivity compensation in another semiconductor, silicon, on the energy of bombarding electrons [67]. It was shown in [67] that the efficiency of silicon compensation grows with the electron energy increasing to 4 MeV and remains unchanged as the energy is raised further to 9 MeV. The rise in the fraction of separated FPs (upon an increase in E_e to approximately 5 MeV) indicates that the average energy of recoil atoms strongly affects (when increasing in the range from 50 to 100 eV) the behavior (separation into components) of primary FPs.

The second factor that accounts for the experimental results [35, 65-68] is the possibility that new, more complex defects can be formed. It has been experimentally demonstrated that point defects are formed in elementary semiconductors (and are not masked by formation of complex defects) under irradiation with electrons having energies of 0.5 to 2 MeV [77, 79]. In [39, 80 - 81], spectra of RDs formed in silicon carbide under irradiation with electrons having various energies were compared. It was shown that, as the electron energy increases to 9 [39] and 15 MeV [81], the defect spectrum is strongly different from that observed under irradiation with 1 MeV electrons. Moreover, traps characteristic of the spectra observed under irradiation with MeV protons appear in the spectra of samples irradiated with electrons having energy of about 15 MeV [35, 81, 82]. This confirms our assumption that the PKA energy is a key factor in the formation of secondary defects. It can be seen in Fig. 2.7 that the values $<E_{PKA}>$ for irradiation with 15 MeV electrons and 1 MeV protons are close to each other.

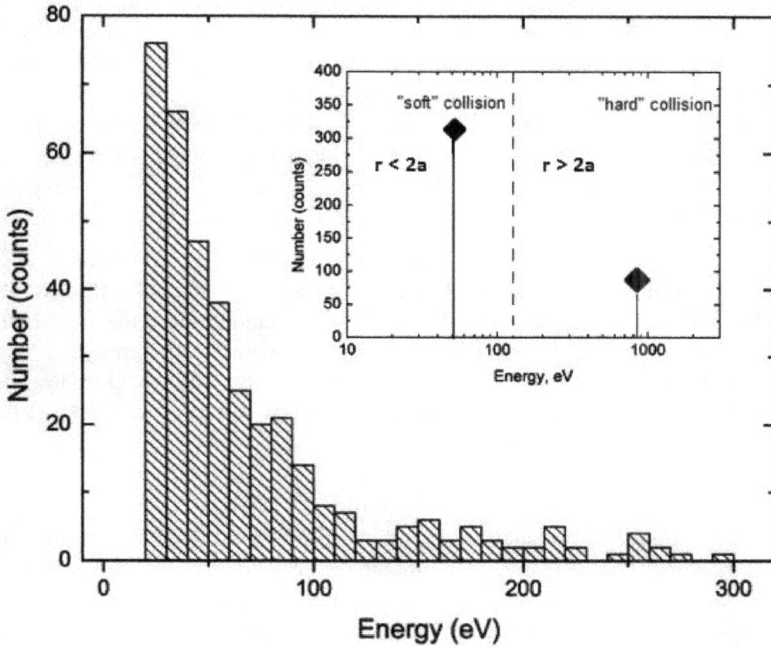

Fig. 2.9. Low-energy part of the histogram for silicon PKAs under irradiation with 8 MeV protons, obtained by calculation with TRIM software [73]. The inset shows the conditional separation of the histogram into the high-energy region of "hard" collisions and low-energy region of "soft" collisions (see text) [84].

Protons with MeV energies can transfer in collisions up to a third of their energy and cause cascades of collisions, with recoil atoms involved in the generation of "secondary" FPs. Using TRIM software [73], we simulated the scattering of 8 MeV protons in silicon. The main goal was to construct a histogram of energies received by primary knocked-on PKA atoms (Fig. 2.9). For clarity, the upper boundary of the spectrum is limited by 300 eV. We distinguish the cases in which the transferred energies are lower and higher than 130 eV. The value of 130 eV corresponds to the maximum energy transferred to a silicon atom by 0.9 MeV electrons (Table 2.7). Of the total 400 collision events to give FPs, 320 events with average energy of primary knocked-on PKA atoms of ~50 eV are present in the left-hand low-energy part of the spectrum (region of "soft" collisions). In the right-hand high-energy part of the spectrum, (region of "hard" collisions), 80 PKAs with average energy of ~840 eV are formed. The share of collisions of this kind is only 20%, but a Si atom with energy of 840 eV creates in a cascade approximately ten vacancies. In contrast to vacancies generated in "soft" collisions similar to the scattering of electrons, these vacancies are not components of single FPs any more. Under these conditions, a cascade yields microscopic regions with sizes of 20 - 30 Å, with up to ten displacements occurring in these regions [68]. Strictly speaking, these microscopic regions are not disordered regions (DRs). However, the high concentration of vacancies in these regions promotes formation of divacancies and complexes of these with impurity atoms [83].

In principle, the irradiation of silicon with MeV protons can be regarded as a superposition of the irradiation with 1 MeV electrons and that with silicon ions with energies of ~1 keV, with the "source" of silicon ions generating these ions uniformly distributed across the sample thickness. Under irradiation with 8 MeV protons, a fifth (20%) of PKAs is in the "hard" part of the spectrum and form displacement cascades. Our estimates demonstrated that, under irradiation with 15 MeV electrons, up to 8% of the total number of recoil atoms may belong to this part of the spectrum.

Conclusion

The effect of the electron irradiation energy on the formation of radiation defects in semiconductors was analyzed. It was shown that the main impact factor is the energy of primary knocked-on atoms (PKAs). As the PKA energy increases, first, the average distance between genetically related Frenkel pairs grows and, as a consequence, the fraction of Frenkel pairs unrecombined under irradiation becomes larger. Second, it becomes possible, with increasing PKA energy, that new, more complex secondary radiation defects can be formed.

References

[1]   Interaction of charged particles with solids and surfaces. Ed. by A.Gras-Marti, H.M.Urbassek, N.R.Arista, F.Flores. (N.Y., Plenum Press, 1991).

[2]   Y.-H.Ohtsuki. Charged Beam Interaction with Solids. (Taylor&Francis Ltd., London and New York, 1983).

[3]   J. W. Corbett and J. C. Bourgoin, in Point Defect in Solids (Plenum, New York, 1975), Vol. 2, p. 1. http://dx.doi.org/10.1007/978-1-4684-0904-8_1

[4]   B. Hudson and B. E. Sheldon, High voltage electron transmission microscopy of pyrolytic silicon carbide coatings from nuclear fuel particles, J. Microsc. 97, 113(1973). http://dx.doi.org/10.1111/j.1365-2818.1973.tb03765.x

[5]   I. A. Honstvet, R. E. Smallman, and P. M. Marquis, A determination of the atomic
        displacement energy in cubic silicon carbide, Philos.Mag. A 41, 201-207 (1980).
        http://dx.doi.org/10.1080/01418618008236135

[6]   H. Inui, H. Mori, and H. Fujita, Electron-irradiation-induced crystalline to
        amorphous transition in α-Sic single crystals, Philos. Mag. B 61, 107- 124,(1990).
        http://dx.doi.org/10.1080/13642819008208655

[7]   W. Jiang, S. Theunthasan, W. J. Weber, and R. Grotzschel, Deuterium channeling
        analysis for He+-implanted 6H–SiC, Nucl. Instrum. Methods Phys. Res. B 161–
        163, 501 (2000). http://dx.doi.org/10.1016/S0168-583X(99)00692-8

[8]   I. Lazanu and S. Lazanu, Long-term damage induced by hadrons in silicon
        detectors for uses at the LHC-accelerator and in space missions, arXiv:hep-
        ph/0209086 (2002).

[9]   A.A.Lebedev, A.M.Ivanov, and N.B.Strokan Radiation Resistance of SiC and
        Nuclear-Radiation Detectors on SiC Films, Semiconductors, 38 (2) (2004) pp 129-
        150

[10]   A.A. Lebedev, V.V. Kozlovski, N.B. Strokan, D.V. Davydov, A.M. Ivanov, A.
        M.Strel'chuk, R.Yakimova Radiation hardness of wide-gap semiconductors (using
        the example of silicon carbide) Semiconductors (2002), V 36, pp 1270-1275
        http://dx.doi.org/10.1134/1.1521229

[11]   A. A. Lebedev, A. I. Veinger, and D. V. Davydov V. V. Kozlovski N. S. Savkina
        and A. M. Strel'chuk Doping of n-type 6H–SiC and 4H–SiC with defects created
        with a proton beam J.Appl.Phys, V 88, (2000),pp 6265-6271
        http://dx.doi.org/10.1063/1.1309055

[12]   A.A. Lebedev, V. V. Kozlovski, S.V. Belov, E.V. Bogdanova, G.A. Oganesyan
        Conductivity compensation in p-6H-SiC in irradiation with 8-MeV protons,
        Semiconductors (2002), V 36, 1145.

[13]   N.S.Savkina, A.A.Lebedev, D.V.Davydov, A.M.Strel'chuk, A.S.Tregubova,
        M.A.Yagovkina, New results in sublimation growth of the SiC epilayers,
        Mat.Science and Eng B61-62, 165 (1999). http://dx.doi.org/10.1016/S0921-
        5107(98)00494-2

[14]   G. A. Lamakina, Yu. A Vodakov, E. N. Mokhov, V. G. Oding and G. F.
        Kholuyanov, "Compared investigation of the electrical properties of the three
        silicon carbide polytypes", Sov. Phys. Solid Stale 12, 2356-2360 (1970).

[15] A.A.Lebedev, E.V.Bogdanova, P.L.Abramov, S.P.Lebedev, D.K.Nel'son, G.A.Oganesyan, A.S.Tregubova, R.Yakimova. Highly doped p-3C-SiC on 6H-SiC substrates, Semiconductor Science Technology 23 (2008) 075004S. http://dx.doi.org/10.1088/0268-1242/23/7/075004

[16] R.A.Smith, Semiconductors, Cambridge University Press, London-New York-Melbourne, 1982.

[17] M.E.Levinshtein, S.L.Rumyantsev, and M.S.Shur, Editors: "Properties of Advanced Semiconductor Materials: GaN, AlN,InN, BN, SiC, SiGe", John Wiley &Sons,Inc.2001

[18] Surface modification and alloying by Laser, Ion, and Electron Beams. Ed. by J.M.Poate, G.Foti, and D.C.Jacobson. (N.Y., Plenum Press, 1983).

[19] H.J. von Bardeleben, J. L. Cantin, I. Vickridge, G. Battistig, Proton-implantation-induced defects in n-type 6H- and 4H−SiC: An electron paramagnetic resonance study, Phys. Rev. B 62 (2000) 10126. http://dx.doi.org/10.1103/PhysRevB.62.10126

[20] U.Littmark, J.F.Ziegler. Handbook of Range Distributions for Energetic Ions in All Eelemens. (N.Y., Pergamon Press, 1980).

[21] Chr.Lehmann. Interaction of Radiation with Solids and Elementary Defect Production. (North-Holland Publishing Compaly, Amsterdam, New York, Oxford, 1977), 341 p.

[22] V.V. Kozlovski, E.V. Bogdanova, V.V. Emtsev, K.V. Emtsev, A.A. Lebedev, V.N. Lomasov.Direct Experimental Comparison of the Effects of Electron Irradiation on the Charge Carrier Removal rate in n-Type Silicon and Silicon Carbide, Mater. Sci. Forum, 483–485, 383 (2005). http://dx.doi.org/10.4028/www.scientific.net/msf.483-485.385

[23] N.Iwamoto, S.Onoda, S.Hishiki, T.Ohshima, M.Murakami, I.Nakano, K.Kawano, Degradation of Charge Collection Efficiency for 6H-SiC Diodes by electron Irradiation, Material Science Forum, 600-603 (2009) 1043). http://dx.doi.org/10.4028/www.scientific.net/MSF.600-603.1043

[24] A.A.Lebedev,V.V.Kozlovskii Irradiation of Sublimation_Grown p_SiC with 0.9_MeV Electrons, Technical Physics Letters, 2014, Vol. 40, No. 8, pp. 651–652 http://dx.doi.org/10.1134/S1063785014080094

[25]   H.Matsuura, N.Minohara and T. Ohsima, Mechanisms of unexpected reduction in
        hole concentration in Al-doped 4H-SiC by 200 keV electron irradiation,
        J.Appl.Phys 104 (2008) 043702 http://dx.doi.org/10.1063/1.2969788

[26]   M. Mikelsen, U. Grossner, J.H. Bleka, E.V. Monakhov, B.G. Svensson, R.
        Yakimova, A. Henry, E. Janzen, A.A. Lebedev. Carrier Removal rate in electron
        Irradiated 4H and 6H SiC Mater. Sci. Forum, 600–603, 425 (2009).
        http://dx.doi.org/10.4028/www.scientific.net/MSF.600-603.425

[27]   R.K.Nadella, M.A. Capano, High-resistance layers in n-type 4H-silicon carbide by
        hydrogen ion implantation, Appl.Phys.Lett, 70, 886 (1997)
        http://dx.doi.org/10.1063/1.118304

[28]   B.G.Svensson, A.Hallen, M.K.Linnarson, A.Yu. Kuznetsov, M.S. Janson,
        D.Adberg, J.Osterman, P.O.A. Person, L.Uultman, L.Storasta, F.U.C. Carlsson,
        J.P.Bergaman, C.Jagadish, E.Morvan.Doping of Silicon Carbide by ion
        implantation, Mater.Sci. Forum 353-356, 549 (2001)
        http://dx.doi.org/10.4028/www.scientific.net/MSF.353-356.549

[29]   A.Hallen, A.Henry, P.Pellegrino, B.G.Svensson, D.Aberg, Ion implantation
        induced defects in epitaxial 4H-SiC, Mater.Sci.Eng. 61-62, 378 (1999)
        http://dx.doi.org/10.1016/S0921-5107(98)00538-8

[30]   G.C.Rybicki, J.Appl.Phys, Deep level defects in alpha particle irradiated 6H
        silicon carbide78, 2996, (1995)

[31]   J. McGarrity, F.McLean, M.Dealancey, J.Palmour, C.Carter, J.Edmond, R.Oakley,
        Silicon Carbide JFET radiation response, IEEE Trans Nucl. Sci 39 1974 (1992)
        http://dx.doi.org/10.1109/23.211393

[32]   N.Camara, K.Zakentes, V.V.Zelenin, P.L.Abramov, A.V.Kirillov, L.P.Romanov,
        N.S.Boltovets, V.A.Krivitsa, A.Thuaine, E.Bano, E.Tsoi, A.A.Lebedev, 4H-SiC
        pn diodes grown by sublimation epitaxy in vacuum (SEV) and their application as
        microwave diodes.Semiconductor Science and Technology, V 23, p. 025016,
        (2008) http://dx.doi.org/10.1088/0268-1242/23/2/025016

[33]   A.A. Lebedev, B.Ya. Ber, G.A.Oganesyan, S.V. Belov, N.V. Seredova, I.P.
        Nikitina, S.P. Lebedev, L.V.Shakhov, V.V. Kozlovski, Effect of 3C-SiC
        Irradiation with 8 MeV protons, to published in Mater.Sci. Forum (2017)

[34]   V. V. Kozlovskii, A. A. Lebedev, and E. V. Bogdanova Model for Conductivity
        Compensation of Moderately Doped n- and p-4H-SiC by High-Energy Electron
        Bombardment, JAP,117 (2015) 155702

[35]   A. Castaldini, A.Cavalini, L.Rigutti, F. Nava, S. Ferrero and F. Giorgis, Deep
        levels by proton and electron irradiation in 4H–SiC, J.Appl.Phys, 98 (2005)
        053706. http://dx.doi.org/10.1063/1.2014941

[36]   V. V. Emtsev, N. V. Abrosimov, V. V. Kozlovski, and G. A. Oganesyan,
        Vacancy-donor pairs and their formation in irradiated n-Si Semiconductors 48,
        1438 (2014). http://dx.doi.org/10.1134/S1063782614110098

[37]   V. V. Kozlovski, V. V. Emtsev, K. V. Emtsev, N. B. Strokan, A. M. Ivanov, V. N.
        Lomasov, G. A. Oganesyan, and A. A. Lebedev, Effect of electron irradiation on
        carrier removal rate in silicon and silicon carbide with 4H modification,
        Semiconductors 42, 242 (2008). http://dx.doi.org/10.1134/S1063782608020231

[38]   A.Castaldini, A. Cavallini, and L. Rigutti, Assessment of the intrinsic nature of
        deep level Z1/Z2 by compensation effects in proton-irradiated 4H-SiC,
        Semicond.Sci.Technol. 21, 724 (2006). http://dx.doi.org/10.1088/0268-
        1242/21/6/002

[39]   L. Storasta, J. P. Bergman, E. Janzén, A. Henry, and J.Lu, Deep levels created by
        low energy electron irradiation in 4H-SiC J. Appl. Phys. 96 (2004) 4909.
        http://dx.doi.org/10.1063/1.1778819

[40]   J.W.Steeds, G.A.Evans, S.Furkert, M.M.Ismail, L.R.Danks, W.Voegeli, and
        F.Carosella, Transmission electron microscope radiation damage of 4H and 6H
        SiC studied by photoluminescence spectroscopy, Diamond and Related Mater. 11,
        1923 (2002). http://dx.doi.org/10.1016/S0925-9635(02)00212-1

[41]   K. Danno, and T.Kimoto. Investigation of deep levels in n-type 4H-SiC epilayers
        irradiated with low-energy electrons J. Appl. Phys. 100, 113728 (2006).
        http://dx.doi.org/10.1063/1.2401658

[42]   H.Kaneko, and T.Kimoto. Formation of a semi-insulating layer in n-type 4H-SiC
        by electron irradiation, Appl. Phys. Lett. 98, 262106 (2011)
        http://dx.doi.org/10.1063/1.3604795

[43]   J.W.Corbett, and J.C.Bourgoin, in Point Defects in Solids, Vol.2. Semiconductors
        and Molecular Crystals, edited by J.H.Crawford and L.M.Slifkin (Plenum Press,
        New York and London, 1975), p.1.

[44]   V.Kozlovski, and V.Abrosimova, Radiation Defect Engineering, Selected Topics
        in Electronics and Systems,Vol.37 (World Scientific, Singapore, 2005).

[45]   Sze, S. M. Physics of Semiconductor Devices, 2nd ed. New York: John Wiley &
        Sons, 1981

[46] V.L.Vinetskii and L.S. Smirnov, About conductivity compensation radiation defects in semiconductors. Sov.Phys. Semicond., 5, (1971) 176.

[47] H.Kaneko and T. Kimoto, Formation of a semi-insulating layers in n-type 4H-SiC by electron irradiation, Appl.Phys.Lett 98, 262106 (2011) http://dx.doi.org/10.1063/1.3604795

[48] In-Tae Bae, W.J.Weber, M. Ishimaru and Y.Hirotsu, Effect of ionization rates on dynamic recovery processes during electron-beam irradiation of 6H-SiC, Appl. Phys. Lett. 90, 121910 (2007). http://dx.doi.org/10.1063/1.2715135

[49] A. A. Lepneva, E. N. Mokhov, V. G. Oding, and A. S. Tregubova, Silicon Carbide irradiated by high doses of neutrons, Sov. Phys. Solid State 33, 1250 (1991).

[50] P. Musumeci, L. Calcagno, M. G. Grimaldi, and G. Foti, Optical defects in ion damaged 6H-silicon carbide Nucl. Instrum. Methods Phys. Res. B 116, 327 (1996). http://dx.doi.org/10.1016/0168-583X(96)00067-5

[51] V. A. Kozlov, V. V. Kozlovski, A. N. Titkov, M. S. Dunaevskii, A. K. Kryzhanovskii., Buried nanoscale damaged layers formed in Si and SiC crystals as a result of high-dose proton implantation, Semiconductors 36, 1227 (2002). http://dx.doi.org/10.1134/1.1521221

[52] E. V. Bogdanova, V. V. Kozlovski, D. S. Rumyantsev, A. A. Volkova, and A. A. Lebedev, Formation and Study of Buried SiC Layers with a High Content of Radiation Defects, Semiconductors, Vol. 38, No. 10, 2004, pp. 1176–1178. http://dx.doi.org/10.1134/1.1808824

[53] A.A.Lebedev, Influence of native defects on polytypism in SiC, Semiconductors 33 (1999) 107. http://dx.doi.org/10.1134/1.1187657

[54] Lebedev A.A., B.Ya. Ber, N.V.Seredova, D.Yu Kazantsev, V.V.Kozlovski, Radiation-stimulated photoluminescence in electron irradiated 4H-SiC, Journal of Physics D: Applied Physics, 48, 485106 (2015). http://dx.doi.org/10.1088/0022-3727/48/48/485106

[55] Y. Koshka and M. Mazzola, Effect of hydrogenation on Al-related photoluminescence in 6H–SiC,Appl. Phys. Lett. 79, 752, (2001). http://dx.doi.org/10.1063/1.1391403

[56] Y. Koshka and M. Mazzola, Photoluminescence Investigation of hydrogen Interaction with Defects in SiC, Mater. Sci. Forum 389–393,609 (2002). http://dx.doi.org/10.4028/www.scientific.net/MSF.389-393.609

[57]   C. Hulsen, N. Achtziger, U. Reinlohner, and W. Witthuhn, Reactivation of Hydrogen-Passivated Aluminum Acceptors in p-type SiC, Mater. Sci. Forum 338–342, 929 (2000). http://dx.doi.org/10.4028/www.scientific.net/MSF.338-342.929

[58]   T. Dalibor, G. Pensl, T. Kimoto, W. J. Choyke, A. Schöner and N. Nordell, Deep Defect Centers in Silicon Carbide Monitored with Deep Level Transient Spectroscopy, Phys. Status Solidi A 162, 199 (1997). http://dx.doi.org/10.1002/1521-396X(199707)162:1<199::AID-PSSA199>3.0.CO;2-0

[59]   A.A.Lebedev "Deep-level defects in SiC material and device sin book: "Silicon Carbide: Material, Proceeding and Devices", ed. by Z.C.Feng and J.H.Zhao, p. 121-163 (2004) Taylor and Francis books Inc.

[60]   A. A. Lebedev, N. A. Sobolev, B.Sh.Urunbaev, Sov. Phys. Semicond. Effect of non-uniform compensation capacitance measurement,16, 1207 (1982)].

[61]   E.V.Astrova, A.A.Lebedev, A.A.Lebedev, Infuence of series resistance of a diode on transient capacitance measurement of deep-level parameters, Sov Phys. Semicond. 19, 850 (1985).

[62]   J. Lutz, H. Schlangenotto, U.Scheuermann, Rik De Doncker. Semiconductor Power Devices Physics. Characteristics, Reliability Springer-Verlag, Berlin Heidelberg, 2011. p.58.

[63]   C.Claeys, E.Simoen. Radiation Effects in advanced semiconductor materials and devices. Springer-Verlag, Berlin Heidelberg, 2002. p.2. http://dx.doi.org/10.1007/978-3-662-04974-7

[64]   Van Allen. Geomagneticically trapped radiation. Space science. John Wiley &Sons. New York. 1963.

[65]   Castaldini A., Cavallini A., Rigutti L., Nava F., Low temperature annealing of electron irradiation induced defects in 4H-SiC, Appl.Phys.Lett. 85, 3780 (2004). http://dx.doi.org/10.1063/1.1810627

[66]   Mikelsen M., Grossner U., Bleka J.H., E.Monakhov, B.G.Svensson, R.Yakimova, A.Henry, E.Janzen, and A.Lebedev, Carrier Removal in Electron Irradiated 4H and 6H SiC, Materials Science Forum. 600-603, 425 (2009). http://dx.doi.org/10.4028/www.scientific.net/MSF.600-603.425

[67]   Wada T., Yasuda K., Ikuta S., Takeda M., and Haruho Masuda. H., Complex defects introduced into Si by high-energy electron irradiation: Production rates of defects in n-Si, J.Appl.Phys. 48, 2145 (1977). http://dx.doi.org/10.1063/1.324032

[68]   Emtsev V.V., Abrosimov N.V., Kozlovski V.V., Oganesyan G.A., Vacancy-donor
        pairs and their formation in irradiated n-Si, Semiconductors. 48,1438 (2014).
        http://dx.doi.org/10.1134/S1063782614110098

[69]   W.A.MacKinley, H.Feshbach, The Coulomb Scattering of Relativistic Electrons
        by Nuclei, Phys.Rev. 74, 1759 (1948). http://dx.doi.org/10.1103/PhysRev.74.1759

[70]   E.Holmström, A. Kuronen, K. Nordlund, Threshold defect production in silicon
        determined by density functional theory molecular dynamics simulations, Phys.
        Rev. B 78, 045202 (2008). http://dx.doi.org/10.1103/PhysRevB.78.045202

[71]   J.W.Steeds, F.Carosella, G.A.Evans, M.M.Ismail, L.R.Danks, W.Voegeli,
        Differentiation between C and Si Related Damage Centres in 4H and 6H SiC by
        Use of 90-300 kV Electron Irradiation Followed by Low Temperature
        Photoluminescence Microscopy, Mater. Sci. Forum 353-356, 381 (2001).
        http://dx.doi.org/10.4028/www.scientific.net/MSF.353-356.381

[72]   Dienes G.J., G.H.Vineyard G.H. Radiation Effects in Solids. New York:
        Interscience Publishers, 1957. P.12.

[73]   Ziegler J.F., Biersack J.P., Littmark U. The Stopping and Range of Ions in Solids.
        New York: Pergamon, 1985. 342 p.

[74]   Kinchin G.H., Pease R.S., The Displacement of Atoms in Solids by Radiation,
        Rep. Prog. Phys., 18, 1 (1955). http://dx.doi.org/10.1088/0034-4885/18/1/301

[75]   A. Belov, A.Mikhaylov, D.Korolev, D.Guseinov, E.Gryaznov, E.Okulich,
        V.Sergeev, I.Antonov, A.Kasatkin, O.Gorshkov, D.Tetelbaum,V.Kozlovski,
        Medium-energy ion-beam simulation of the effect of ionizing radiation and
        displacement damage on SiO2-based memristive nanostructures, Nucl.Instr. Meth.
        in Phys. Research, B 379, 13 (2016)

[76]   A.M.Ivanov, I.N.Ilyashenko, N.B.Strokan, B.Schmidt, Recombination of
        nonequilibrium carriers in the tracks of heavy ions in Si1, Semiconductors. 29, 43
        (1995).

[77]   J.W.Poate and J.S.Williams, in Ion Implantation and Beam Processing, edited by
        J.S.Williams and J.M.Poate. Sydney: Academic,1984. P.13.
        http://dx.doi.org/10.1016/B978-0-12-756980-2.50006-0

[78]   B. G . Svensson, C. Jagadish and J. S . Williams, Generation rate of point defects
        in silicon irradiated by MeV ions, Nucl.Instr. Meth. in Phys. Research B 80/81,
        583 (1993).

[79] B.G.Svensson, J.L.Lindstrom, Generation of divacancies in silicon by MeV electrons: Dose rate dependence and influence of Sn and P, J.Appl.Phys. 72, 5616 (1992). http://dx.doi.org/10.1063/1.351961

[80] S.A.Reshanov, S.Beljakowa, B.Zippelius, G.Pensl, K.Danno, G.Alfieri, T.Kimoto, S.Onodo, T.Ohshima, F.Yan., R.P. Devaty., W.J.Choyke, Thermal Stability of Defect Centers in n- and p- type 4H-SiC Epilayers Generated by Irradiation with High-energy Electrons, Materials Science Forum. 645-648, 423 (2010). http://dx.doi.org/10.4028/www.scientific.net/MSF.645-648.423

[81] G.Alfieri, E.V.Monakhov, B.G.Svensson, A.Hallen, Defect energy levels in hydrogen-implanted and electron-irradiated n-type 4H silicon carbide, J.Appl.Phys. 98,113524 (2005). http://dx.doi.org/10.1063/1.2139831

[82] A. M. Strel`chuk, V. V. Kozlovski, A. A. Lebedev, N. S. Savkina, D. V. Davydov, V. V. Solov`ev, M.G.Rastegaeva, Doping of 6H–SiC pn structures by proton irradiation, Nucl.Instr.Meth. in Phys. Research B ., 147, 74 (1999). http://dx.doi.org/10.1016/S0168-583X(98)00581-3

[83] R.Poirier, V.Avalos, S.Dannefaer, F.Schiettekatte, S.Roorda, Divacancies in proton irradiated silicon: Comparison of annealing mechanisms studied with infrared spectroscopy and positron annihilation, Nucl. Inst.Meth. in Phys. Research B 206, 85 (2003).

[84] V.V.Kozlovski, A.A.Lebedev, V.V.Emtsev, G.A.Oganesyan, Effect of the Energy of Recoil Atoms on Conductivity Compensation in Moderately Doped n-Si and n-SiC under Irradiation with MeV Electrons and Protons, Nuclear Instruments and Methods in Physics Research Section B 384 (2016) 100-105. http://dx.doi.org/10.1016/j.nimb.2016.08.003

CHAPTER 3

Radiation defects in SiC and their influence on recombination processes

Abstract

Results from current studies of the parameters of radiation defects arising in 6H-, 4H-, and 3C-SiC after irradiation by electrons, protons, neutrons and some kind of ions are analyzed. Presented data on the ionization energy and capture cross sections of centers formed by doping SiC with different types of charge particles or during irradiation, as well as of intrinsic defects. The involvement of these centers in radiative and non-radiative recombination is examined. The photoluminescence (PL) arising in n- and p-type 4H-SiC upon electron irradiation (0.9 MeV) has been analyzed. A conclusion is made that the PL is activated by donor—acceptor pairs constituted by a nitrogen atom and a structural defect.

Keywords

Electrons, Protons, Neutrons, Ions, Intrinsic Defects, Lifetime Killer, Photoluminescence

Contents

3.1   Introduction

The history of the development of SiC growth technology and devices on its basis can be described as a fight against the existing crystal lattice defects, both macroscopic (micropipes, dislocations, and inclusions of other polytypes) and microscopic (point defects, background impurities, and complexes of these). The structural defects still affect parameters of SiC devices, such as the mobility of carriers, their lifetime, and diffusion length.

It has been found previously that the Si/C ratio varies between SiC polytypes: it decreases with increasing degree of hexagonality, being 1.046, 1.022, and 1.001 for 3C, 6H, and 4H polytypes, respectively [1]. The data on diffusion and solubility of impurities in various silicon carbide polytypes, considered in [2], also pointed to different concentrations of carbon vacancies (V_C). The explanation suggested by the authors for this dependence is that, as the lattice strain grows with increasing concentration of carbon vacancies, bonds between atoms in the cubic sites become more energetically favorable. And just this circumstance leads to restructuring of the crystal and transformation of the polytype. The authors of [2] also believed that most parts of V_C are in an electrically inactive state.

The assumption that carbon vacancies are the main complex-forming defect of SiC was also supported by the data of various authors, which suggested that V_C are involved in the formation of various centers and separate vacancies merge into stable clusters [3, 4]. Also, a good correlation was noted between the decrease in the concentration of carbon vacancies with increasing percentage of hexagonality of the polytype and decreasing background concentration of intrinsic defects in SiC [5].

The studies carried out during the last 10 - 15 years have confirmed the previous assumption that V_C play an important part in the formation of defects and defect complexes in SiC and affect parameters of devices based on silicon carbide.

3.2   Intrinsic defects in silicon carbide

Because intrinsic defects in the crystal lattice of SiC and the related deep levels in the energy gap are frequently present in samples before their irradiation, or the concentrations of these defects grow in the course of irradiation, we believe that it is appropriate to begin the consideration of radiation defects by describing the most known intrinsic defects.

3.2.1   Centers in the lower half of the energy gap

L-center. Deep centers (DCs) with ionization energy $E_v + 0.24$ eV and $\sigma_p \approx 10^{-15}$ cm^2 (L-centers) have been found in studies of 6H-SiC p-n structures produced by ion implantation of Al (IL structures) [6]. The L-center was also found in p-n structures produced by sublimation epitaxy (SE structures). In the implanted structures, its concentration increased near the metallurgical boundary of the p-n junction; in samples of other types, there was no noticeable distribution profile of L-centers.

The ionization energy of the L-center was close to the ionization energy of Al, furnished by other methods [7]. Later, analogs of L-centers were found in 4H-SiC produced by sublimation [8] and 4H- and 6H-SiC formed by the CVD method [9, 10]. It was shown in [7] that, as the concentration of Al is raised from 10^{18} to 10^{21} cm^{-3}, its ionization energy decreases from 0.27 to 0.1 eV. The presence of "shallow" and "deep" aluminum has also been found in other SiC polytypes [10, 11]. It has been shown that there exist two separate centers, rather than a single center with two charge states. The structure of these centers is not fully understood. Presumably, "deep" aluminum has the form of a complex of an Al atom and a structural defect.

i-center. In addition to L-centers, a deep center in the lower half of the energy gap, i-center, has been observed in the DLTS spectra of 6H-SiC IL structures [6]. The ionization energy of the i-center was found to be within the range $E_v + (0.52 - 0.58)$ eV. Structures similar to IL structures have been obtained on the basis of epitaxial 4H-SiC films. Their DLTS spectrum was similar to the spectra of 6H-SiC IL samples, and an analog of an i-center with $E_v + 0.53$ eV was found in these samples [9, 12]. The fact that the parameters of i-centers in 6H and 4H are close indicates that structures of these centers are similar for the given polytypes. The distribution of i-centers in the base region of a diode has been measured in IL structures based on 6H- and 4H-SiC [8]. The extrapolation of the distribution profile of i-centers in 4H-SiC structures yielded a good coincidence with the beginning of the profile in structures based on 6H-SiC, which suggests that the distributions of i-centers in structures based on 4H- and 6H-SiC coincide in nature. A comparison of the distribution profiles of i-centers and the $N_d - N_a$ profile obtained from C-U characteristics suggested that the distribution of compensating defects in the base coincides with the distribution of i-centers. Thus, the compensation of the base region in IL structures is due to the increased concentration of deep acceptor centers (i-centers) near the metallurgical boundary of the p-n junction. The acceptor nature of the i-center is also confirmed by the carrier capture cross-section ratio ($\sigma_p \gg \sigma_n$) for this DC. It was also found in [12] that the implantation of Al into the purest epitaxial 4H-SiC layers gave rise to an S-shaped current-voltage characteristic, which was attributed to the high concentration of i-centers.

In [12], the activation of the so-called "defect photoluminescence" (DPL) was attributed to the formation of i-centers (see Section 3.5 for more detail). The fact that the DPL appears in SiC polytypes upon their irradiation with various kinds of charged particles gives reason to assume that the i-center is a complex mostly composed of intrinsic defects of the crystal lattice of SiC, the concentration of which grows upon irradiation or implantation. This assumption is in good agreement with the results of [13], where a conclusion that a carbon vacancy is contained in the center activating the luminescence was made on the basis of experiments on the thermal stability of defect electroluminescence (DEL) in SiC crystals.

***D*-center.** Although the *D*-center is not a purely structural defect (it presumably contains a boron atom), it is a characteristic background center in 6H- and 4H-SiC grown by various methods [6, 14, 15, 16 -19]. This center was first observed in a study of DLTS spectra in SiC<B> structures and, together with i-centers, in a study of SE samples [6]. The fact that the values of ionization energy and σ_p are close for i- and *D*-centers resulted in a superposition of their DLTS peaks in SE and IL structures. Therefore, the ionization energy of DCs (E^{av}) determined for structures of these types fluctuated within the range 0.52 eV $< E_i <$ 0.58 eV. A study of parameters of i- and *D*-centers at electric field strengths in the range (1-7) 10^5 V/cm did not reveal any noticeable dependence of the ionization energies of these deep levels (DLs) on the electric field strength [20].

Although the ionization energy of the *D*-center exceeds that of the i-center by more than 10%, the hole capture cross-sections show the opposite relationship, with the result that the recharging time constants of these centers are rather close at T $\approx$ 300 K. The deeper *D*-center is recharged at lower temperatures, and, therefore, the overall DLTS peak looks like a peak from a single center (Fig. 3.1). The i-DLTS method has been used to divide the signals from the DLs under consideration [21]. This method has a high resolving capacity because of yielding, in fact, the first derivative of a DLTS spectrum. In addition, the application of other temporal windows made it possible to move to the temperature range in which the recharging time constants differ more strongly for the DLs being observed. Figure 3.1 shows an i-DLTS spectrum of an SE structure, in which the signals from the originally existing centers are well seen [20].

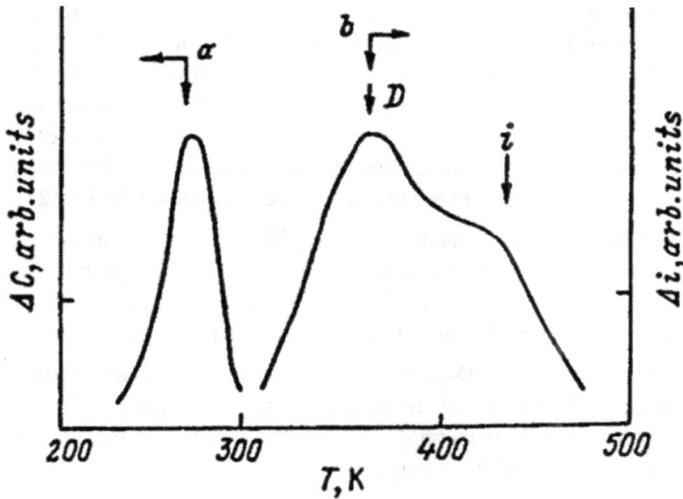

Fig. 3.1. (a) DLTS and (b) i-DLTS spectra of an SE p-n structure with i- and D-centers. Parameters of the spectra: (a) $t_1 = 10$ ms and $t_2 = 30$ ms; (b) $t_1 = 0.1$ ms and $t_2 = 5$ ms. Adapted from [20].

Thus, according to [6, 20], both i- and D-centers exist in SE structures, and the ionization energies of these centers are independent of the electric field strength in the space-charge layer in the ranges of concentrations and reverse voltages under study. An application of the method for resolution of DLTS signals to the structures under study demonstrated that D-centers are predominant in SiC<B>, and i-centers, in IL structures. An additional diffusion of boron into epitaxial ES layers before the p-n junction was formed also led to predominance of D-centers. An analog of the D-center ($E_v + 0.54$ eV) has been observed in epitaxial 4H-SiC layers doped with boron in the course of growth [22].

The D-center is occasionally named "deep boron" because a shallower acceptor center ($E_v + 0.35$ eV) exists in SiC<B>samples in addition to the D-center. The formation of two kinds of levels that are not different charge states of the same center upon diffusion of boron into SiC is apparently due to the complex nature of the diffusion distribution of B in SiC. For example, it was suggested in [13] that boron atoms diffuse as associates with intrinsic defects of the silicon carbide lattice. Later studies of the diffusion of boron from implanted layers confirmed this point of view [14, 16-18]. According to [17, 18], B diffuses into SiC by the "kick-out" mechanism via silicon interstitial sites, as it occurs when boron diffuses in silicon. It was shown that the diffusion mostly yields D-centers, whereas mostly "shallow boron" is formed in doping with boron during the growth of epitaxial layers. On the whole, it seems quite probable that the surface branch of the diffusion distribution is due to shallow-boron centers, i.e., to boron atoms occupying sites of the SiC crystal lattice. At the same time, the bulk branch is formed by boron +intrinsic defect complexes. This standpoint has been confirmed by later studies [23, 24].

According to ESR data, a boron atom substituting Si (Bsi) corresponds to the structure of the shallow boron center, and deep boron is a complex of a boron atom and a carbon vacancy [25, 26]. It is noteworthy that the D-center is a level characteristic of SiC and it has been found in the material grown by various methods [27].

It is also noteworthy that the D-center is an activator of the characteristic photo- and electroluminescence in silicon carbide polytypes [28, 29], similarly to "deep aluminum" (L-center) [30 and references therein]. The spectral position of the peak of these PL depends on the energy gap width of a given polytype.

Other defects. A center near the midgap (E_v + 1.41 eV) was observed in [9] in p-6H-SiC and related to a intrinsic defect. SiC samples doped with Mn, V and irradiated with neutrons were studied in [31]. A comparison of PL spectra and ESR data demonstrated a deep acceptor associated with the appearance of a red PL. A study of lightly doped (N_d - $N_a \approx 10^{14}$ cm^{-3}) n- and p-type 6H-SiC crystals grown by the CVD method revealed a set of levels whose concentrations increased as the C/Si ratio became larger [32]. D-centers were predominant in the p-type material; in addition, traps with energies E_v + 0.64 eV and E_v + 1.3 eV were observed [33].

An acceptor DL with energy E_c - 1.5 eV was found in a study of n-4H-SiC [34]. A study of p-4H-SiC revealed a deep donor (E_v + 1.49 eV) identified by the authors as a single carbon vacancy (HK4 center).

3.2.2 Defects in the upper half of the energy gap

6H-SiC

R and S centers. A study of 6H-SiC Schottky diodes revealed two DCs, S center (E_c - 0.35 eV, $\sigma_p \approx 10^{-15}$ cm^2) and R center (E_c - 1.27 eV) (Fig. 3.2) [35]. The concentrations of these levels, N_R and N_S, coincided to within 10-20% in all the samples under study. The values of N_R and N_S ($\sim 10^{15}$ cm^{-3}) were close to those in IL and SE p-n structures produced on the basis of these epitaxial layers. Thus, it can be concluded that no significant change in the concentrations of R and S centers occurred in fabrication of the p-n junction. It was shown in [36] that these centers are formed upon irradiation and ion implantation of 6H-SiC.

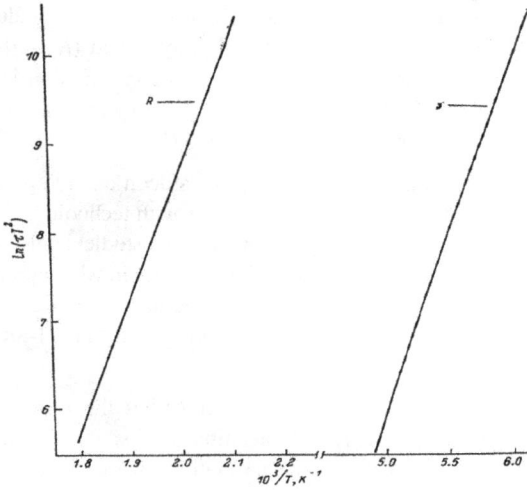

Fig. 3.2. Arrhenius dependences for the S and R centers [35].

A double center $(E_1 \backslash E_2)$ $(E_c - 0.34 \backslash E_c -0.41$ eV) close in parameters to the S center was observed first in Lely substrates [37] and then also in epitaxial layers grown by the CVD method [38]. In [35], R and S centers were regarded as the main nonradiative recombination centers (see Section 3.4). The structure of these centers was identified as that of a complex of silicon vacancies on the basis of ESR data [39].

Z1/Z2. One more double peak Z_1/Z_2 $(E_c - 0.6....0.7$ eV) was observed in [37]. Later, this peak was also found in CVD-grown epitaxial layers [40]. Earlier, the ESR method was employed to study several deep centers in 6H-SiC Lely substrates [41]. One of these centers with energy of ~600 meV was associated with the V_c-Vsi divacancy. In [42], an assumption was made that this center corresponds to the Z_1/Z_2 center found by the DLTS method.

71

4H-SiC

Z_1 ($Z_{1/2}$). The background level Z_l with energy E_c - (0.63-0.68) eV is characteristic of 4H SiC. This level has been observed in materials grown by various technological methods [22, 43-45]. According to [43], the concentration of Z_l falls with decreasing value of N_d - N_a in an epitaxial layer. However, it is still present in lightly doped ($N_d \approx 10^{14}$ cm^{-3}) CVD-grown epitaxial layers. At the same value of N_d - N_a, its concentration is lower in layers with high dislocation density. A rise in the concentration of the $Z_{1/2}$ defect with increasing oxygen of the nitrogen impurity was observed in [44].

It is known that 4H-SiC has the best parameters among the silicon carbide polytypes as regards the fabrication of semiconductor devices. After the growth technology of 4H-SiC substrates and epitaxial layers was mastered, a large number of studies concerned with structural defects in this materials were carried out, including those in which properties of the Z1 center were analyzed [45-50]. The interest in the properties of this center became even more profound after it was found that it is a charge carrier killer in 4H-SiC [51-56] (see Section 3.4).

Various authors assumed that the structure of the $Z_{1/2}$ center has the form of various combinations of structural defects - vacancies and interstitial atoms. The problem of the structure of the given center was solved by using the method of irradiation with low-energy electrons [45]. Different amounts of energy are required for defects to be formed in silicon and carbon sublattices in SiC. If electrons with <200 keV energies are used, defects are only formed in the carbon sublattice. Thus the rise in the concentration of the $Z_{1/2}$ defect with increasing dose of irradiation with low-energy electrons demonstrated that this center definitely contains V_C. A comparison of the results of DLTS and ESR measurements on the same 4H-SiC samples subjected to different irradiation doss demonstrated that the center is a single carbon vacancy in the charge state (2-/0) [49]. This conclusion agrees with the results obtained in reducing the concentration of the $Z_{1/2}$ defect in the course of oxidation [47], or implantation of C, Al. or Si [57]. These technological processes result in that an excess amount of C atoms appears in SiC, and these atoms recombine with V_C thereby making lower the concentration of $Z_{1/2}$.

$EH_{6/7}$. A deeper center (E_c - 1.5 eV) named $EH_{6/7}$ has been observed in 4H-SiC samples in addition to the I center [45, 48, 58]. It has also been shown that the structure of the $EH_{6/7}$ center contains a carbon vacancy [45, 48] and the given center is an acceptor [59].

In [60], an interesting analogy was drawn between the E_1/E_2 and R centers in 6H-SiC and $Z_{1/2}$ and $EH_{6/7}$ centers in 4H-SiC. Both pairs of centers are intrinsic defects in SiC, their concentrations increase upon irradiation and implantation, all these centers are not annealed out up to temperatures of ~1500 °C and exhibit similar types of high-temperature annealing. In addition, the concentration of all the four centers substantially decreased upon oxidation (Fig. 3.3).

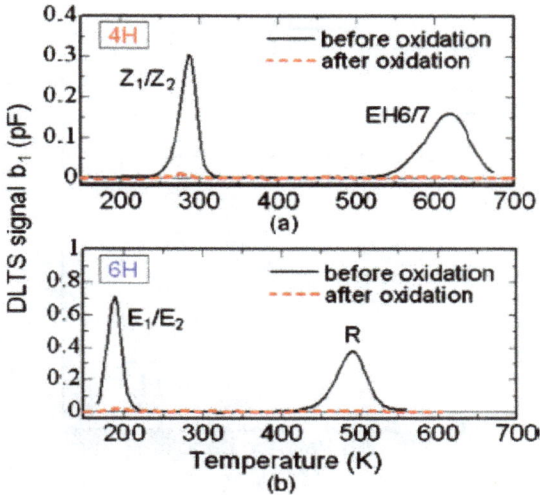

Fig. 3.3 DLTS spectra of n-type (a) 4H-SiC and (b) 6H-SiC before and after the thermal oxidation at 1150 °C for 6 h. Before being oxidized, the samples were irradiated with electrons (energy 150 keV, fluence 2 × 10^{17} cm^{-2}) and annealed at 950 °C [60].

The authors of [60] suggested that, similarly to the $Z_{1/2}$ and EH$_{6/7}$ levels in 4H-SiC, the E_1/E_2 and R levels in 6H-SiC correspond to two charge states of a single carbon vacancy. In this case, the position of the levels relative to the conduction band bottom will be determined by the empirical Langer-Heinrich (LH) rule [61], which has been previously suggested for III-V and II-VI compounds and successfully used to describe Ti levels in SiC [22]. According to LH, the energy position of the levels related to this center is invariable among isovalent semiconductor compounds relative to a certain common energy level (e.g., valence band top). In other words, the mutual positions of levels in SiC polytypes will remain constant relative to the E_v level and be independent of the parameters of a polytype. At the same time, the energy positions of the levels relative to E_c will be determined by the energy gap width of the given polytype (see Fig. 3.4).

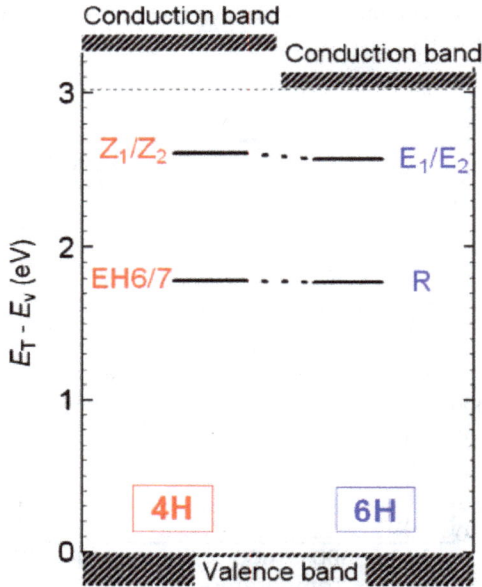

Fig. 3.4. Energy position in the energy gap of the thermally stable traps E_1/E_2 in n-6H-SiC and $Z_{1/2}$ and $EH_{6/7}$ in n-4H-SiC [60].

Background defects in 3C-SiC were studied in epitaxial layers grown on silicon substrates [62, 63]. In [62], two centers were observed (SC_1 E_c - 0.34 eV and SC_2 E_c - 0.58 eV), with their DLTS peaks being hardly visible on the background of the complete relaxation band. Bands of this kind could be due to the high density of bulk defects (dislocations etc.). In eight years, the quality of epitaxial films was substantially improved and this was reflected upon the DLTS spectrum reported in [63], in which signals from separate levels are already well distinguishable. The authors of [63] observed three DCs [T_1 (E_c - 0.32 eV), T_2 (E_c- 0.52 eV), and T_3 (E_c - 0.56 eV)], with the shallowest of which being identified with the previously found SC_1.

Only a single DL (E_c - 0.62 eV), close in parameters to T_3 and SC_3, was found in 3C-SiC epitaxial layers grown by sublimation epitaxy on a 6H-SiC substrate [64]. In [65], the thermal stability of intrinsic defects in 3C-SiC crystals commercially manufactured by HOYA company was studied. A shallower level (E_c - 0.14 eV) annealed at temperatures of ~1500°C was found in addition to two DCs that are close in parameters to SC_1 and SC_2. Based on the activation energy of annealing and on the known theoretical models, the authors of [65] suggested that this DL is associated with a silicon vacancy.

Parameters of main intrinsic defects in SiC are shown in tables 3.1 and 3.2.

Table 3.1 Parameters of main structural defects in 6H SiC.

Centre	Energy position, eV	structure	Participation in recombination processes
S ,(E$_1$/E$_2$)	Ec – 0,45	Vc	
R	Ec – 1,25	Vc	Life time killer
D ("deep boron")	Ev + 0,58	B + Vc (?)	CA* recombination hv ~ 2,1 eV
i –centre	Ev + 0,5	?	DAP** recombination hv ~ 2,3 eV
L ("deep aluminium")	Ev + 0,24	Al + Vc (?)	DAP and CA recombination hv ~ 2,5 – 2,7 eV

Table 3.2 Parameters of main structural defects in 4H SiC.

Centre	Energy position, eV	structure	Participation in recombination processes
Z $_{1/2}$	Ec – 0,68	Vc	Life time killer
E $_{6/7}$	Ec – 1,48	Vc	
D ("deep boron")	Ev + 0,56	B + Vc (?)	CA recombination hv ~ 2,3 eV
i –centre	Ev + 0,5	?	DAP recombination hv ~ 2,5 eV
"deep aluminium"	Ev + 0,23	Al + Vc (?)	DAP and CA recombination hv ~ 2,7 – 2,9

*) Recombination between C-zone and acceptor level

**) Donor- acceptor pair recombination

3.3 Radiation doping of SiC

It is known that the energy released in deceleration of a particle or an ion in a semiconductor results in that atoms are shifted from their equilibrium positions, i.e., structural (radiation) defects are formed. Being disruptions of the periodic structure of a crystal, these defects create allowed states within the energy gap of the semiconductor (deep levels). This section briefly reviews the results obtained in studies of parameters of deep levels that appear in SiC under irradiation with various particles.

3.3.1 Electrons

6H-SiC. A study of n-6H-SiC irradiated with 3.5 - 4 MeV electrons revealed defects in the upper half of the energy gap, with energies of 0.35, 0.6, and 1.1 eV [66]. All these defects are eliminated by annealing below temperatures of ~1300 K. According to their ionization energies, these defects can be identified as native defects: S center, Z_1/Z_2, and center at Ec - 1.06 eV. In a later study, these experiments were repeated by using irradiation with 2 MeV electrons [67]. In addition to an increase in the concentration of the background defects E_1/E_2 and Z_1/Z_2, a new center was observed: E_3/E_4 at Ec - 0.57 eV. A similar increase in the concentration of the S center (E_1/E_2) was observed for CVD-grown epitaxial layers upon their irradiation with 2 MeV electrons [38]. In addition, a center was observed at Ec - 0.51 eV, which was eliminated by annealing already at temperatures of ~800 °C. In [68], n-6H-SiC epitaxial layers were irradiated with electrons having energies of 0.2, 0.3, 0.5, and 1.7 MeV. The appearance of radiation defects was recorded only upon irradiation with electrons having energies of $\geq$ 0.5 MeV. In addition to the structural defects E_1/E_2 and Z_1/Z_2, traps with energies Ec - 0.23, Ec - 0.32, and Ec - 0.5 eV appeared, which were eliminated by annealing even at temperatures of 300 °C.

Two traps were observed at Ev + 0.36 and Ev + 0.81 eV upon low-energy (0.4 MeV) irradiation of p-6H-SiC with electrons, with these defects being eliminated by annealing at 500 and 350 °C, respectively [69]. A new center appeared at Ev + 0.74 eV upon annealing at 500 °C and disappeared upon annealing at 900 °C. Two more traps were formed at Ev + 0.56 and Ev + 0.81 eV upon annealing at 700 °C and disappeared upon annealing at 900 °C. A center at Ev + 0.71 eV appeared upon annealing at 1400 °C and was preserved up to annealing temperatures of 1600 °C. At this temperature, all the discovered centers were fully eliminated by annealing. Upon irradiation with 1 MeV electrons, two centers were observed, in addition to the D center, at Ev + 0.52 and Ev + 0.24 eV (presumably i and L-centers), which were stable up to being annealed at 1800 °C, and also centers at Ev + 0.48 and Ev + 1.2 eV, which were eliminated by annealing at 400 and 1000 °C, respectively. Upon irradiation of p-6H-SiC with 1.7 MeV electrons, two deep centers were found at Ev + 0.55 and Ev + 0.78 eV. Both these DCs were eliminated at temperatures of 200-500 °C.

4H-SiC. As already noted in the previous section, the structure of Z_1 and EH_6/EH_7 centers was identified by using a low-energy electron irradiation in which defects are only created in the carbon sublattice. In [45], an increase in the concentration of only Z_1 and EH_6/EH_7 centers was observed upon irradiation with 116 keV electrons. These centers were eliminated by annealing only at temperatures of 1600-1700 °C. A conclusion was made on the basis of the results obtained that these centers have a similar structure and contain a carbon vacancy. After n-4H-SiC CVD structures were irradiated with electrons having energies in the range 2-2.5 MeV. The formation of quite a number of defects was observed in addition to an increase in the concentration of the background level Z_1: EH_1 (Ec - 0.45 eV), EH_2 (Ec - 0.68 eV), EH_4 (Ec - 0.72 eV), EH_5 (Ec - 1.15 eV), and EH_6/EH_7 (Ec - 1.65 eV) [21, 71-73]. Most of these centers also appeared upon implantation of He and some other ions [22]. In [22], several measurement procedures were used to examine in detail the EH_6/EH_7 center. It was shown that the center is constituted by two levels with close ionization energies. It is noteworthy that the paired values of the ionization energies are a characteristic feature of SiC, which is apparently due to the existence of several nonequivalent positions in its crystal lattice. In [74, 75], the effect of irradiation with 6 and 8.6 MeV electrons, respectively, was studied. In [74], the influence exerted by the technological conditions in which the epitaxial layers were grown (growth rate, C/Si ratio, and nitrogen concentration) on the subsequent formation of Z_1 and EH_6/EH_7 centers under irradiation was additionally analyzed. In [75], annealing of some radiation defects (Ec - 0.39 eV) at temperatures of 60 – 140 °C and an increase in the concentration of the structural defect Z_1 at annealing temperatures of 100 – 170 °C was observed by analogy with 6H-SiC. An EPR study of the irradiated material demonstrated that the signal associated with the carbon vacancy is preserved up to annealing temperatures of ≥ 1600 °C [76].

The properties of radiation defects in p-4H-SiC were also studied by irradiation with low-energy (160 - 400 keV) electrons [77-79]. In [77], several DCs, including the D center, were observed in as-grown samples. A reactive ion milling of the samples, followed by annealing at 1150 °C, gave rise to two centers at Ev + 0.79 and Ev + 0.84 eV. Upon irradiation with 160 - 400 keV electrons, two new centers appeared in all the samples at Ev + 0.98 and Ev + 1.44 eV. After the annealing at 950 °C, these centers were eliminated and the concentration of the center at Ev + 1.44 eV increased. In the case of an electron irradiation with energies exceeding 160 keV, followed by annealing at 950 °C, centers with energies Ev + 0.3, Ev + 0.58, and Ev + 1.44 eV were observed. All the DCs, with the exception of the D center, were eliminated by annealing at 1550 °C. The authors of

[77] concluded that all the DCs they observed are complexes that include a structural defect associated with the displacement of a carbon atom. In [79], several centers were observed at energies Ev + 0.21....0.29 eV upon irradiation. After p-4H-SiC was irradiated with 4.6 MeV electrons, the concentration of the DC at Ev + 0.2 eV, attributed by the authors to an Al impurity [80], decreased. The concentration of another center of unknown nature at Ev + 0.2 eV remained unchanged. In [81], p-4H-SiC irradiated with 2.5 MeV electrons was studied by the method of electron spin resonance with photoexcitation (photo-ESR). The photo-ESR data obtained for the positively charged carbon vacancy (V_C^+) can be accounted for in terms of the model of a deep donor with charge state (+/0), which gives rise to a level with energy of (1.47 ± 0.06) eV.

3.3.2 Neutrons

A number of DCS were also observed upon irradiation of SiC with neutrons [82, 83] (Ec -- 0.5, Ec - 0.24, and Ec - 0.13 eV). It was shown that p-SiC subjected to neutron irradiation exhibits a week electronic conduction before being annealed [84]. A high-dose neutron irradiation ($\sim 10^{21}$ cm^2) led to amorphization of 6H-SiC and 15R-SiC samples. An annealing of the amorphized samples led to formation of inclusions of the 3C-SiC polytype [85]. Several later studies concerned with the effect of neutron irradiation on the properties of SiC are known [86, 87]. Mostly the influence exerted by the irradiation of this kind on the current-voltage characteristics of devices was analyzed. It was reported that the carrier removal rate for SiC is about 4.5/carrier/cm^3 /neutron/cm^2, which is approximately three times smaller than the value for silicon [88]. A deep center at Ec - 0.49 eV was observed upon irradiation of 3C-SiC. This center was eliminated by annealing at 350°C [89]. The carrier removal rate was found to be 7.2 cm^{-1}, which is close to the value of 7.8 cm^{-1}, obtained for Si irradiated with the same neutron spectrum. According to the results of an EPR study of the irradiated and annealed 6H-SiC [90], only the model of a defect constituted by an interstitial carbon and a carbon vacancy (C_{Si}-V_C) in the twice positively charged state can account for all the experimental results. It was suggested in [90] that this defect is formed from an isolated silicon vacancy that appears under annealing via displacement of a neighboring carbon atom toward the vacancy. A conclusion was made in [91] that a complex defect stable at high temperatures appears in 6H-SiC heavily irradiated with neutrons and annealed at 1500 °C. It was suggested that this complex may include four vacancies V_{Si}-$3V_C$ bound to interstitial atoms $(C2)_{Si}$ or to a pair of two interstices $(C2)_{Si}$-Si_C.

Separate mention can be made of studies in which a homogeneously doped SiC was obtained via its irradiation with thermal neutrons (see [92] and references therein). In the course of irradiation, the following nuclear reaction occurs:

$$^{30}\text{Si} + \text{n} \rightarrow {}^{31}\text{Si} \rightarrow (\beta^- \text{ decay}, T_{1/2} = 2.62 \text{ h}) \, {}^{31}\text{P}.$$

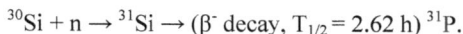

where $T_{1/2}$ is the half-life of ^{31}Si.

The radiation defects introduced in the process are further eliminated by annealing. This technology is used to fabricate silicon-based power devices. One of problems encountered in its application is the low content of the ^{30}Si isotope (~3%) in natural silicon. In [92], silicon carbide was grown from a source enriched with the ^{30}Si isotope to ~99%. As result, Nd - Na concentrations of ~10^{17} cm^{-3} were reached. Also, the EPR signal from donor phosphorus levels was studied in detail in relation to the annealing temperature of the samples.

3.3.3 Alpha - particles

It was reported in [40] that irradiation of n- and p-type 6H-SiC with α-particles leads only to an increase in the concentration of the already existing background defects. A conclusion was made that the radiation hardness of SiC compares well with that of InP, another radiation-hard semiconductor. The effect of α-particles on n-4H-SiC was studied in [46]. It was found that the background structural defect Z_1, which is a carbon vacancy, was transformed into a Z_1/Z_2 center, which contains a silicon atom in addition to V_C.

3.3.4 Protons

In [93], n-4H-SiC was irradiated with 1.2 MeV protons and 15 MeV electrons. On being irradiated, the samples were annealed at temperature in the range 100-1200 °C with a step of 50 °C. In addition to those known from the literature, three new RDs were observed in proton-irradiated samples: Ec - 0.69, EC - 0.73, and Ec - 1.03 eV. Having compared the results obtained in experimental studies of the distribution profiles of these centers with those furnished by theoretical calculations of the distribution of the implantation-induced defects, the authors of [93] came to a conclusion that all the defects observed in the study contain hydrogen atoms.

The DCs formed in n-type 4H-SiC and 6H-SiC upon irradiation with 8 MeV protons were studied in detail in [94-97]. The parameters of these centers were close to those of the centers previously observed in electron-irradiated SiC (Tables 3.3 and 3.4). A comparison with the EPR data demonstrated that the centers are either carbon vacancies [Ec - 0.5 eV (6H-SiC); Ec - 0.63-0.7 eV (4H-SiC)] or a pair of vacancies in the carbon and silicon sublattices. The different thermal ionization energies at the same supposed structure ($V_{Si} + V_C$) can be presumably accounted for by the different distances between the components (vacancies) of the pairs, characteristic of each of these RDs. Similar results have been obtained in a study by positron spectroscopy of 6H-SiC irradiated with low-energy protons [98]. It was found that the irradiation yields various types of $V_{Si} + V_C$ divacancies and also separate vacancies eliminated by annealing at temperatures of ~900 °C [98]. It was also assumed that one of the divacancies is responsible for the DC near the midgap [97]. At the same time, it was reported in [99] that only silicon monovacancies eliminated by annealing at 1100 °C were found in n-6H- and 4H-SiC irradiated with 12 MeV protons. A study by positron spectroscopy of 6H-SiC originally doped with aluminum and nitrogen and irradiated with 12 MeV protons demonstrated that vacancy-related defects were also present in the samples prior to irradiation [100]. A negatively charged silicon vacancy was observed in the samples before and after the irradiation. The neutral divacancy $V_{Si} + V_C$ was formed upon irradiation and served as a predominant

Table 3.3. RDs observed in 6H-SiC after different type of irradiation.

n-type 6H-SiC					
Energy position of observed RDs, , eV	Electron irradiation [38,66-68, 165]	Neutron Irradiation [82-88]	Proton irradiation [94-101, 107]	Structural defects	Possible structure
Ec – (0.16-0.2)	+	+	+		Primary defects [107
Ec -0.36/0.4	+	+	+	E₁/E₂ [44] S [35	Vsi-complex [39]
Ec -0.5	+	+	+		Vc [66,107]
Ec -0.7	+	+	+	Z₁/Z₂ [44]	Vc + Vsi [44,107]
Ec -0.8	+		+		
Ec – (1.1 – 1.22)	+		+	R [35]	Vc + Vsi [70,90]
p-type 6H-SiC					
Energy position of observed RDs, eV	Electron irradiation [69]	Structural defects		Structural defects	
Ev + 0.24	+	L –centre		Al + Vc (?) [6-11]	
Ev + 0.36	+				
Ev+ 0.56	+	D-centre		B + Vc (?) [6,14-21]	
Ev + 0.71	+				
Ev + 0.81	+				
Ev + 1,2	+				

positron trap at room temperature. In addition, various types of ionic defects were detected. One of these acts as a strong trapping center even at room temperature. In [101], data obtained in studies of n-6H-SiC irradiated with protons having various energies were analyzed. It was shown that R-centers (Ec - 1.1 eV) have the highest introduction rates of 0.17, 70, and 700 cm^{-1} for, respectively, protons with energies of 1 GeV, 8 MeV, and 150 keV.

Table 3.4. RDs observed in 4H-SiC after different type of irradiation

n-type 4H-SiC					
Energy position of observed RDs, , eV	Electron irradiation [21-22,45 71-76,]	a-particles [46]	Proton irradiation [93-99, 107]	Structural defects	Possible structure
Ec – 0,18	+		+	E$_c$-0.18÷0.2 [43]	Primary defects [107]
Ec – 0.45	+		+		Vacancy+impurity[70]
Ec – (0.63-0,7)	+	+	+	Z$_1$ [44]	Vc [56,107]
Ec -0.96	+		+		Vc + Vsi [107]
Ec -1.0	+		+	Ec-1.1 eV [43]	Vc + ? [45,48]
Ec – 1.5	+		+	EH$_{6/7}$ [45,48,58]	
p-type 4H-SiC					
Energy position of observed RDs, eV	Electron irradiation [77-81]	Proton irradiation [94-98]	Structural defects	Possible structure	
Ev + 0,2....0.29	+	+	L-centre	Al + ? [80]	
Ev + 0.55	+	+	D-Centre	B + Vc (?) [6,14-21]	
Ev + 0.63...0.7	+	+		Vc + Vsi [98]	
Ev + 0.79	+	+			
Ev + 0.98	+	+			
Ev + 1.44	+	+		Vc [81]	

The possibility of performing a protonic (hydrogen) passivation of silicon carbide has been studied [102 - 106]. It was shown that the room-temperature resistivity of n-4H-SiC exceeded 8×10^6 Ω cm upon irradiation with 350 keV protons at a dose of 1×10^{14} cm^{-2} [106]. As the sample was heated, its resistance rapidly decreased. Irradiation with 8 MeV protons also led to an increase in the dc electrical resistance (Rb) of SiC structures at room temperature. In contrast to 6H-SiC, 4H-SiC exhibited a decrease, rather than an increase, in the total concentration of uncompensated donors in proton-irradiated samples. This shows that, under irradiation, acceptor centers were formed in the lower half of the energy gap, or donor centers were disintegrated in its upper half. In addition, the irradiation resulted in that deep acceptor centers were formed, to which electrons passed from shallower donor levels.

Under heating, Rb decreased exponentially with activation energy ϵ_A (Fig. 3.5). With increasing irradiation dose, ϵ_A increased and asymptotically approached the value of ~1.1 eV for 6H-SiC and 1.25 eV for 4H-SiC [107]. This resulted in that n-6H-SiC layers semi-insulating at room temperature were obtained. Layers of this kind can be used in the technology of devices that are not intended for operation at high temperatures, e.g., photodetectors or various radiation detectors.

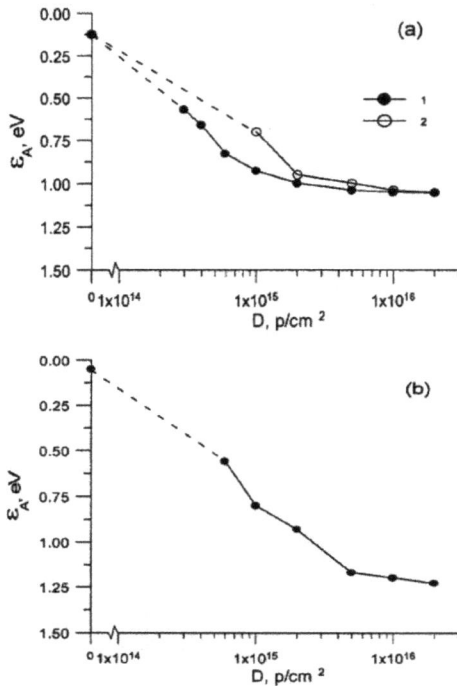

Fig. 3.5. (A) 6H-SiC, ϵ_A vs. irradiation dose for (1) CREE p-n structure and (2) Schottky diode based on an epitaxial layer grown by sublimation epitaxy. (B) 4H-SiC, ϵ_A vs. irradiation dose for a Schottky diode based on an epitaxial layer grown by sublimation epitaxy [107].

In [108], the effect of irradiation with 5 MeV protons (at a dose of 10^{11} cm^{-2}) on 4H-SiC p-n diodes was studied. The presence of a p-n junction enabled the authors to examine RDs formed in both the lower and upper halves of the energy gap of n-6H-SiC. Ten traps were found, which mostly coincided in their parameters with the previously observed RDs and structural defects.

3.3.5 Ion implantation

The ion-implantation technology is rather widely used in development of device structures based on silicon carbide. Doping with nitrogen and aluminum is used to create heavily doped regions in n- and p-type SiC, respectively. The implantation of argon is used to create high-resistivity regions on the periphery of the working regions of Schottky diodes. There also exist more exotic applications of the ion implantation method, e.g., for creating SiC-AlN solid solutions. Apparently, the deceleration of ions gives rise to defects in the crystal lattice. In the post-implantation annealing, these defects interact with impurity atoms to give defect complexes. Previously, we already mentioned some of the studies concerned with RDs in SiC samples subjected to implantation. Below, we consider several recent relevant publications.

In [109, 110], the DLTS method was used to study RD parameters upon implantation of N, P, and Al ions into n- and p-4H-SiC. Also, the effect of the post-implantation annealing on the parameters of the deep centers found in the study was examined. $Z_{1/2}$ and EH$_{6/7}$ centers were predominant in the as-grown 4H-SiC. After the implantation and annealing at 1000 °C, at least seven new RDs were observed. Only $Z_{1/2}$ and EH$_{6/7}$ centers remained upon a high-temperature annealing at 1700 °C. In the p-type material, D and HK 4 (Ev + 1.4 eV) centers were predominant prior to implantation. After the implantation and annealing at 1000°C, two more levels were recorded at Ev + 0.35 and Ev + 0.7 eV. After the annealing at 1700°C, two more RDs were found at Ev + 0.72 and Ev + 1.3 eV in all p-type samples, in addition to D and HK 4. Depth analyses have revealed that the major deep levels are generated at larger depth as compared with the implantation profile. In [110], the same research team studied the influence exerted by oxidation on the concentrations of the most thermally stable RDs appearing upon implantation of N, P, and Al ions. It was shown that the oxidation of n-4H-SiC at 1150 °C strongly reduced the concentrations of $Z_{1/2}$ and EH$_{6/7}$. However, the HK 4 center was preserved in p-4H-SiC and a new RD (IP4) appeared at Ev + 0.72 eV. The concentration of IP4 could be reduced by annealing of the samples in the atmosphere of argon at 1400 °C. These phenomena can be accounted for in terms of the model in which excess interstitials are generated at the interface being oxidized and diffuse into the bulk region.

In [111], germanium-implanted n-4H-SiC samples were studied. It has been shown previously that the introduction of this isovalent impurity in the course of the CVD growth of SiC leads to a decrease in the resistance of the epitaxial layers due to the rise in the carrier mobility [112]. It was found that the joint implantation of Ge and Ar leads to a higher concentration of structural defects in SiC. It was unexpectedly found that an increase in the concentration of ^{74}Ge resulted in a decrease in the concentration of $Z_{1/2}$. In this case, a DLTS signal was recorded from a new RD with ionization energy Ec - 0.79 eV. This decrease in the concentration of the $Z_{1/2}$ center was not observed upon implantation of Ar, which creates disruptions in the SiC crystal lattice that are similar to those in implantation of Ge. The authors of [111] suggested that Ge atoms create local stresses in the crystal lattice of silicon carbide and these stresses serve as getters for carbon vacancies.

In a recent publication [113], the method of electrically detected magnetic resonance was used to study silicon carbide samples upon ion implantation of nitrogen atoms. The presence of a stable complex Nc-Vsi (nitrogen atom at carbon site + silicon vacancy) was found experimentally and confirmed by a theoretical simulation.

In [114], the temperature dependence of the amorphization process of 3C-SiC in implantation of argon ions. It was found that this process has a number of specific features at temperatures exceeding 200 °C. The authors attributed these features to the high-temperature interaction of the defects being formed.

It can be noted in the end of this section that the observed diversity of parameters of the centers formed in silicon carbide under irradiation and subsequent annealing may be due to the following two reasons. First, this may be due to the difference in the concentration and type of background impurities contained prior to irradiation in samples grown in various technological installations. The interaction of these impurity atoms with structural defects formed under irradiation creates complexes that may have various ionization energies and carrier-capture cross-sections. Second, the presence of several nonequivalent positions for atoms in the crystal lattice of SiC may give rise to centers with close parameters, i.e., to the "paired nature" characteristic of silicon carbide. At various ratios between the concentrations of these centers, the total energy found from, e.g., a DLTS spectrum may vary. It should be noted that the problem of the influence exerted by the nonequivalent positions in the SiC lattice on the ionization energies of impurity and defect centers has not been studied to the full extent so far.

Despite the certain diversity of the results presented in this section, two important conclusions can be made.

(1) The most characteristic structural defect of SiC is the carbon vacancy V_C, with which the main recombination centers S ($E_{1/2}$) and R in 6H-SiC and $Z_{1/2}$ and $EH_{6/7}$ in 4H-SiC are associated. The concentrations of V_C and the related DCs grow upon all kinds of irradiation.

(2) Upon irradiation of SiC at room temperature, the annealing and reconstruction of the RDs being formed begins at temperatures of ~500 °C. All kinds of RDs are fully eliminated by annealing at 1600°C and above. Hence follows the conclusion that the RDs formed in SiC at room temperature are primary radiation defects.

3.4   Radiation defects and nonradiation recombination in SiC

It has been found previously [115, 116] that the lifetime of holes in n-SiC falls within the range from 10^{-7} to 10^{-9} s, and their diffusion length, within the range 0.01-1 μm. Apparently, the main recombination in SiC as an indirect-gap semiconductor involves deep levels (DLs). However, no centers that could determine such a short lifetime have been found.

In [117, 118], the temperature dependence of the diffusion length of holes in n-6H and 4H-SiC layers produced by various technological methods was studied. It was found that the diffusion length (Lp) grows with increasing temperature. This dependence Lp = f(T) can be accounted for either by the involvement in the recombination of sufficiently shallow levels, or by the negative temperature dependence of the cross-section of carrier capture by a deep recombination center.

In [119], an analysis of current-voltage characteristics and values of Lp in 6H-SiC p-n structures led to a conclusion that the main nonradiative recombination involves multiple-charge centers whose parameters vary between structures produced by different technologies. However, these centers were not found in experiments. It was suggested in [120] that vanadium is the main nonradiative recombination center in SiC. This conclusion was based on the inversely proportional relationship between the intensity of Al-N DAP recombination and the intensity of luminescence associated with the intracenter transition in the vanadium center. This means that the intensity of the donor-acceptor recombination is quenched as the concentration of V increases. In our opinion, this proof is not definitive because the DAP recombination intensity is affected not only by the carrier lifetime, but also by a large number of widely diverse factors. These include, among other things, the concentration of centers included in a DAP, which was not determined in any way by the authors of [120].

In [35, 121], the recombination characteristics of minority carriers and DC parameters in 6H-SiC p-n structures grown by various methods were subjected to a comprehensive analysis. A processing of the experimental data demonstrated that the relaxation lifetime of minority carriers (τ_{rel}) is within the range 2-30 ns. The diffusion length of minority carriers for SE p-n structures with various background doping levels was found to be 1.5 - 0.05 µm.

The supposed lifetime was calculated with consideration for the parameters of the DLs observed. The results obtained in lifetime calculations demonstrated that the only level whose parameters can account for the experimentally observed lifetime is the S DL. An approximately an order of magnitude larger value of τ_{rel} was obtained in a calculation made with consideration for the parameters of the R-center, the concentrations of the R and S centers being the same in structures of both types. An analysis of how τ_{rel} and L_p^2 depend at room temperature on the concentration of S and R DLs demonstrated that these quantities are in an inversely proportional relationship.

The experimentally observed temperature dependence of Lp could be only accounted by the parameters of the shallower S-center. It has been suggested that a double-charge R-S center can exist, which determines the recombination parameters of both kinds of 6H-SiC p-n structures under study. After the CVD technique for growth of epitaxial SiC layers was fully developed, the recombination characteristics were also studied in samples of this type [122, 123]. The results of a kinetic study of the low-temperature PL were used to estimate the carrier lifetime at 0.45 µs, which substantially exceeds the values in SiC epitaxial layers grown by other methods. Also, S and R DLs were found in CVD-grown 6H-SiC layers, with their concentration being one to two orders of magnitude lower than that in SE structures [22].

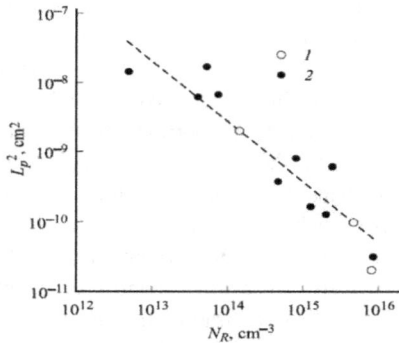

Fig. 3.6. Dependence of the squared diffusion length of holes in 6H-SiC samples on the concentration of the R-center: (1) after irradiation and annealing, (2) unirradiated [96].

Figure 3.6 shows the dependence of Lp on the concentration of S and R DLs, taken from [96] and supplemented with the data we obtained in studies of CVD-grown samples and the highest purity ($\sim 10^{15}$ cm^{-3}) SE samples. It can be seen from the figure that the inversely proportional dependence of Lp on the concentration of the levels is preserved for all three kinds of samples. Thus, these centers are presently the most probable candidates for the role of "lifetime killers" in 6H-SiC.

It was found for 4H-SiC that L_P and τ_p strongly depend on the injection level and vary from 2 μm and 15 ns (low injection level) to 6-10 μm and 140-400 ns (high injection level), respectively [124]. Later, it was shown that τ_p values obtained by various measurement methods may be strongly different. A possible reason is the thin layer at the interface between the lightly doped N-base of the diode and the heavily doped P+ emitter. This layer can accumulate structural defects that appear because of the lattice mismatch between the lightly and heavily doped SiC layers. This leads to substantially smaller values of τ_p in this layer, compared with the rest of the base. Depending on the measurement procedure, the values obtained for τ_p may correspond to this layer, or to the bulk of the base region [125]. It has been shown that the τ_p values better corresponding to the switching characteristics of devices can be obtained by using the procedure suggested by B. Gossik [126]. The values presently obtained for epitaxial SiC layers produced by the CVD method exceed 10^{-6} s [127]. It was found that annealing of n-4H-SiC samples at a temperature of 2600 °C leads to an increase in the hole lifetime from <10 ns to 3 ms. The authors of [127] found that the lifetime increases simultaneously with the decrease in the concentration of the $Z_{1/2}$ DL, which made it possible to recognize this center as the main "lifetime killer" in 4H-SiC. This conclusion was confirmed by various studies carried out with various measurement procedures [128, 129]. In [49, 50, 130, 131], experiments with implantation of carbon into 4H-SiC led to a conclusion that the $Z_{1/2}$ DL is associated with the carbon vacancy in the crystal lattice in 4H-SiC. Thus, it became clear that the lifetime can be made longer via introduction into SiC of atoms that recombine with V_C and thereby reduce the concentration of $Z_{1/2}$ centers. This conclusion was confirmed in experiments with oxidation of SiC [48] and implantation of C, Si, and Al ions [57, 132]. It has also been shown that, when present in concentrations lower than 10^{13} cm^{-3}, the $Z_{1/2}$ center ceases to affect the experimentally measured lifetime (which corresponds to $\tau_p \approx 10^{-6}$-10^{-5} s) (Fig. 3.7) [133]. Presumably, the key influence is exerted under the given conditions by recombination processes involving other DLs, or by the surface recombination.

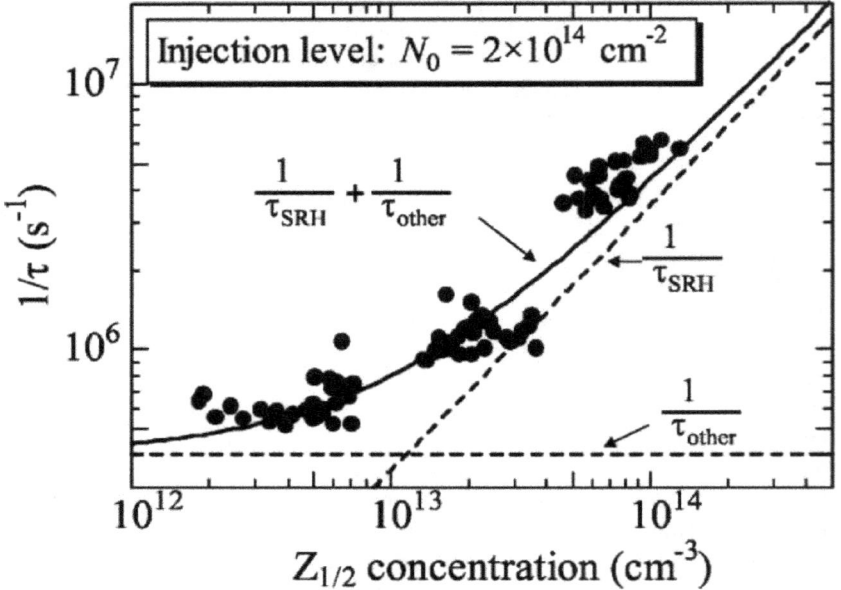

Fig. 3.7. Dependence of the lifetime on the concentration of $Z_{1/2}$ centers in 4H-SiC [133].

To conclude this section, it can be noted the main recombination centers in 4H- and 6H-SiC were identified and their structure and methods for both reducing and raising their concentration were determined, which is highly important for development of devices based on silicon carbide. Obtaining significant diffusion lengths of minority carriers made it possible to modulate the resistance of power SiC devices with thick base regions, and also to improve the charge collection (sensitivity) of UV photodetectors and nuclear detectors. The observed increase in the concentration of recombination centers in the course of SiC irradiation enables control over τ_p and even fabrication of regions with different lifetimes in the same device.

3.5   Radiation – stimulated photoluminescence in SiC

3.5.1   "Defect" photoluminescence

The short-wavelength luminescence in the energy range 2.6-2.3 eV was discovered in 1966 [135] in n-SiC (6H) crystals irradiated with K and Li ions and then annealed. The luminescence spectrum was constituted by two triplets of narrow lines (H- and L-lines) situated near energies of 2.6 eV and a broad structureless band peaked at 2.35 eV. Later, it was found that this photoluminescence (PL) appears after SiC is irradiated with electrons or neutrons and also may be present in the spectrum of unirradiated samples (see review [30] and references therein). It was found in [136,137] that the broad band is not formed via development of a fine structure, being due to the radiative recombination involving the donor level of nitrogen and an acceptor center that appears in the course of implantation. In [138], the structure of the H - and L-lines and their temperature dependences were studied in detail and the spectrum itself was named "D1 spectrum."

The D1 spectrum was recorded in SiC upon its irradiation with electrons [140], neutrons [141], and various kinds of ions [136], which made it possible to form on the basis of 6H-SiC ion-doped with Al and Ga effective LEDs emitting light in the green spectral range.

In other SiC polytypes, the irradiation gave rise to luminescence with similar properties [136, 144] (Figure 3.8). In this case, the PL peak position was shifted relative to that in 6H-SiC to shorter wavelengths (for the broader bandgap polytype) or to longer wavelength (for the narrower gap polytype) by an amount approximately equal to the difference between the energy gap widths of a given polytype and 6H-SiC. Because a luminescence of this kind appears as a result of irradiation or upon introduction of various kinds of ions into SiC, it was assumed that the center activating the luminescence either has a purely defect structure [144], or is a complex constituted by a native defect and an atom of a background impurity [145]. However, no deep centers (DCs) associated with this defect were observed.

Later, a certain terminological confusion occurred in the opinion of the authors [12], because only H- and L-lines and their phonon replicas were understood as the D1 spectrum in some reports [137, 138, 145], and in others [146, 147], this name was extended to include the whole spectrum of this luminescence. As a result, part of authors attributed to this mechanism the whole spectrum observed in [135] after it was shown that the H- and L-lines can be accounted for by the recombination of a bound exciton. In the opinion of the authors of [12], this is incorrect because of coming into contradiction with the earlier obtained results [136-137].

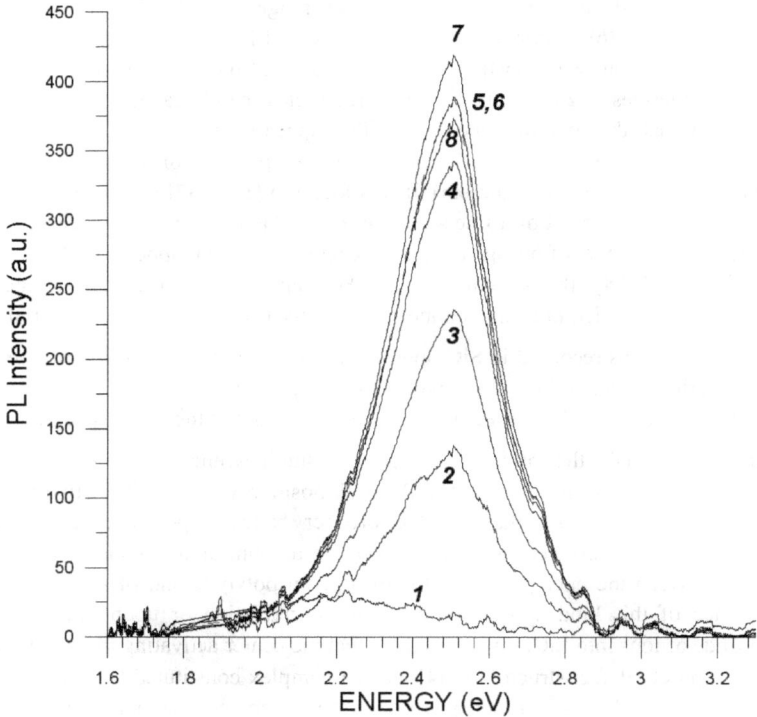

Fig. 3.8. PL spectra of n-4H-SiC before (curve 1) and after electron irradiation. The curves correspond to different irradiation doses (the curve number is followed by the corresponding irradiation dose in parentheses in units of 10^{17} cm^{-2}): 2 (1), 3 (1.5), 4 (2.5), 5 (3.5), 6 (5), 7 (6.5), and 8 (8) [148].

It was suggested in [12] to leave the term "D1 spectrum" for the short-wavelength part of the spectrum observed in [135] (H- and L-lines) and to name the long-wavelength part of the given spectrum (broad structureless band) the "defect-related" photoluminescence (DPL).

Possibly, both the DPL and D1 spectrum are due to the carrier recombination involving DCs of similar nature (or even DCs of the same type); however, particular emission mechanisms may be different for both parts of the spectrum.

An analysis of the DPL characteristics in terms of various recombination mechanisms was made in [12]. It was found that the DPL peak shifts with increasing forward current J to shorter wavelengths, the DPL relaxation is nonexponential, and the DPL peak shifts to longer wavelengths as the recording time increases after the injection pulse ends. All the above specific features of the DPL are characteristic indications of the donor-acceptor recombination mechanism [149].

The analysis made in [12] led the authors to conclude that the main recombination occurs via a DAP constituted by an i-center and a nitrogen level. Calculations performed in terms of this model with consideration for the parameters of the centers and p-n structures were in good agreement with the experiment, which, however, leaves open the possibility that a bound exciton is formed at the i-center. The optical lines due to this exciton will lie in another spectral range. In addition, the presence of nonequivalent positions in the crystal lattice of SiC for both donor and acceptor centers may result in the formation of several DAPs with close parameters, each of these contributing to the overall spectrum.

The spectral position of the DPL peak can also be accounted for by the radiative transition of an electron from the conduction band to an i-center with consideration for the Frank-Condon effect. However, this model fails to explain he strong quenching of the DPL at below-room temperatures. This is so because, according to the depth at which the level related to the i-center lies, the noticeable thermal ionization of this center will begin at temperatures higher than 300 K. Thus, the dependences of the DPL intensity on the excitation level and temperature, considered in [12], also supported the model suggested in [136-137], according to which the recombination occurs via an i-center-nitrogen DAP.

The DPL characteristics were further studied in [148]. The dependence of the DPL intensity on the electron irradiation dose was examined experimentally and theoretically in samples with various nitrogen concentrations. The 4H-SiC silicon carbide samples under study had the form of 50-μm-thick epitaxial films grown by the CVD method. The samples were grown on conducting substrates with (0001)Si orientation, manufactured by CREE company. The concentrations of uncompensated acceptor (donor) concentrations Na-Nd (Nd-Na) in the unirradiated samples were found from capacitance-voltage (C-U) characteristics. The irradiation with 0.9 MeV electrons was performed on a resonant transformer accelerator (pulse repetition frequency 490 Hz, pulse width 330 μs, duty cycle ~6) on a target cooled with running water. The electron beam was extracted into air through an exit window closed with a thin (~50 μm) titanium foil. Under these conditions, the monochromaticity of the electron beam was not worse than 10%. The average electron beam current density was 12.5 μA cm^{-2}. The irradiation dose was measured by the method of Faraday's local cylinders. The range of 0.9 MeV electrons is 1.0 mm in SiC. The defects can be considered to be introduced under irradiation uniformly throughout the sample volume because the thickness of the SiC samples being irradiated was substantially smaller than the range of the charged particles.

The PL was excited with a nitrogen laser operating at a wavelength of 337.1 nm and having the following parameters: pulse power 2 kW, pulse width 10 ns, pulse repetition frequency 100 Hz. The pumping power density was ~50 kW/cm^2. The PL spectra were measured at liquid nitrogen temperature (77 K). The spectra were studied for the samples both in their unirradiated state and after each irradiation dose.

Table 3.5. Parameters used in the calculation

Parameters	Values		
	Sample no. 1	Sample no. 2	Sample no. 3
Concentration (conductivity type)	2-3 10^{14} (P)	2 10^{15} (P)	1.1 10^{16} (N)
N_{MAX}, cm^{-3}	0.5 10^{16}	1.5 10^{16}	5 10^{16}
I_0, rel. units	30	8	7
I_{MAX}, rel. units	60	100	185
γ, cm^2	2.35 10^{-15}	2.35 10^{-15}	0.6 10^{-17}
α, cm^3	12 10^{-15}	6.7 10^{-15}	3.7 10^{-15}
N^0_{DAP}, cm^{-3}	0.25 10^{16}	0.18 10^{16}	0.19 10^{16}
N_0, cm^{-3}	0.25 10^{16}	1.32 10^{16}	4.81 10^{16}

After the dependence of the DPL intensity on the irradiation dose was measured, the concentration of nitrogen atoms was determined in the same samples by dynamic secondary-ion mass spectrometry using CAMECA IMS 7f ion microprobe. A beam of ^{133}Cs$^+$ primary ions with impact energy of 15 keV was scanned over a 100×100 μm^2 area, with secondary negative molecular ions ^{12}C^{14}N$^-$ collected from a central area 60 μm in diameter. The mass spectrometer was tuned to a mass resolution M/ΔM ≈ 7500 to separate the $^{12}C^{14}N^-$ analytical signal from the mass interference from $^{13}C^{2-}$ secondaries. During the SIMS analysis, the specimens were irradiated with a low-energy electron beam to avoid charging. The sputtered-crater depths were determined with an AMBIOS XP-1 calibrated stylus profile meter. The SIMS data were quantized by using relative-sensitivity factors [150], with a nitrogen-implanted standard SiC sample. The experimentally measured nitrogen concentrations are listed in Table 3.5.

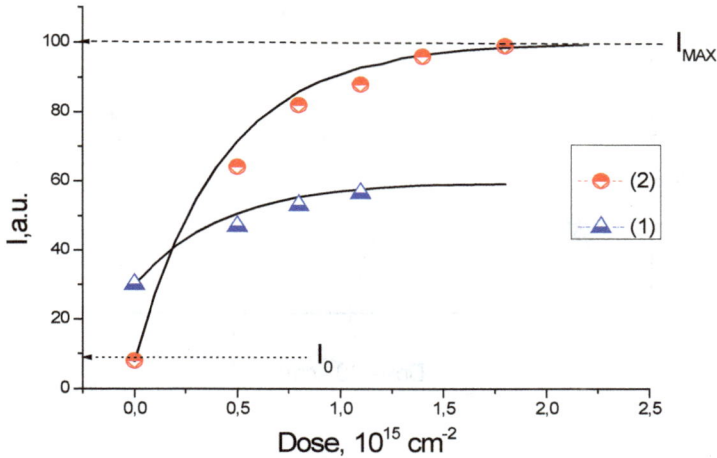

Fig. 3.9. DPL intensity I for p-4H-SiC vs. the electron irradiation dose: (1) data for sample no. 1 and (2) data for sample no. 2. The lines represent the calculation by formula (3.4) with the parameters from Table 3.5 [148].

Prior to irradiation of the samples, the PL spectrum contained a broad band peaked at hv ≈ 2.2 eV, commonly attributed to the radiative recombination via acceptor levels of boron (D-center) [30] (Fig. 3.8). Apparently, its presence is due to the uncontrolled doping of the as-grown epitaxial layers with boron. After the irradiation, the DPL band peaked at hv ≈ 2.5 eV started to appear in the PL spectrum. It has been shown previously that irradiation of p-4H-SiC samples results in the appearance of a DPL whose intensity levels-off ("saturates") with increasing irradiation dose [152] (Fig. 3.9). In [148], the dependence of the DPL intensity on the irradiation level [I = F(D)] was examined in n-4H-SiC samples. It was shown that, as also in the case of p-4H-SiC, the dependence I = F(D) also levels-off, but does so at a larger irradiation dose (Fig. 3.10).

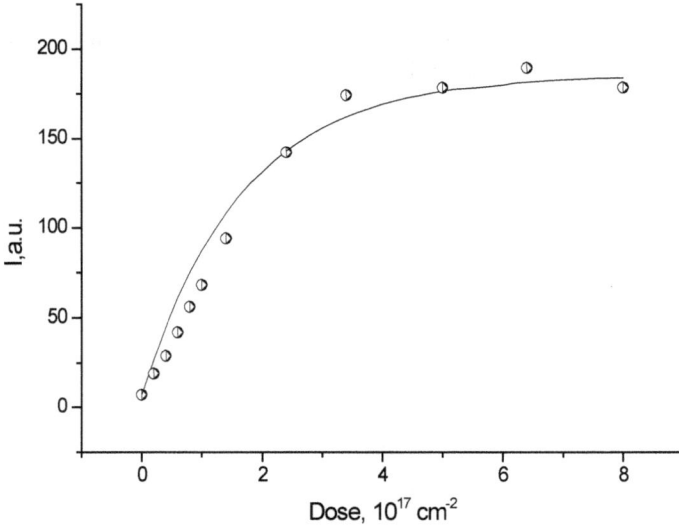

Fig. 3.10. DPL intensity for n-4H-SiC vs. the electron irradiation dose (sample no. 3). The line represents the calculation by formula (3.4) with the parameters from Table 3.5. [148]

The expression for the DPL intensity I can be written as

$$I = \alpha(N^0_{DAP} + N_{DAP}) \tag{3.1}$$

where α is the coefficient that relates the external excitation intensity to the DPL intensity, N^0_{DAP} is the initial concentration of the donor-acceptor pairs activating the DPL, and N_{DAP} is the concentration of the donor-acceptor pairs formed upon irradiation.

In [34,151], various mechanisms of the radiation-induced compensation of semiconductors were analyzed. By analogy with [34, 151], we can write an expression describing the variation of the DAP concentration in the course of irradiation as

$$dN_{DAP}/dt = n_{FP} \beta \, G \, \tau \, N \tag{3.2}$$

Here, $N = N_0 - N_{DAP}$; N_0 is the concentration of nitrogen atoms at the beginning of irradiation, which are not included in DAPs; n_{FP} is the generation rate of Frenkel pairs under irradiation; β is the constant of interaction between a nitrogen atom and a radiation defect (dependent on the velocity of defect motion over the crystal and on the probability of interaction upon encounter); G is the flux of electrons; and τ is the lifetime of a radiation defect.

Equation (3.2) has the following solution under the boundary conditions

$$t = 0 \Rightarrow N_{DAP} = 0 \text{ and } t = \infty \Rightarrow N_{DAP} = N_0:$$

$$N_{DAP} = N_0[1 - \exp(-n_{FP} \beta G \tau t)] \tag{3.3}$$

Taking into account that $G t = D$, where D is the irradiation dose, and replacing $n_{FP} \beta \tau$ with γ, we substitute (3.3) in (3.1):

$$I = \alpha N^0_{DAP} + \alpha N_0[1 - \exp(-\gamma D)] \tag{3.4}$$

It is easy to extract $I_0 = \alpha N^0_{DAP}$, which is the PL intensity before the beginning of irradiation, and $I_{MAX} = \alpha N^0_{DAP} + \alpha N_0$, i.e., the value at which the intensity levels-off, from the experimental dependence of the PL intensity on the irradiation dose (Fig. 3.10). Physically, this means that nearly all nitrogen atoms are incorporated into DAPs.

Then, equation (3.4) can be easily transformed to

$$\ln[(I_{MAX} - I)/(I_{MAX} - I_0)] = -\gamma D \tag{3.5}$$

From the slope ratio of the dependence described by expression (3.5), we can find the value of γ (Table 3.5). A graphical representation of the experimental dependence (3.5) is shown in Figure 3.11. It can be seen that the dependences nearly coincide for p-type samples, but, at the same time, noticeably differ for n-SiC.

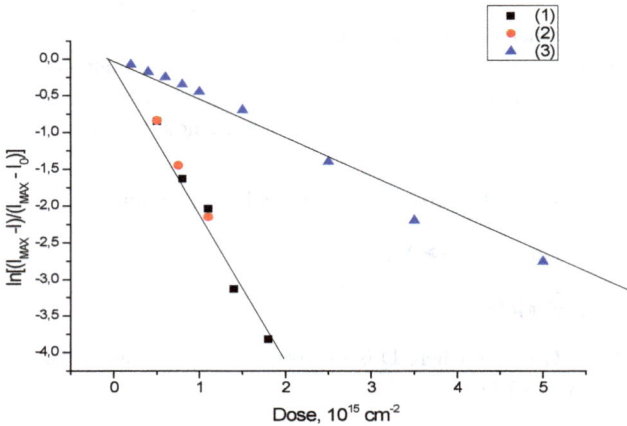

Fig. 3.11. Dependence of ln [(I_{MAX}-I) \ (I_{MAX}-I_0)] on the dose of electron irradiation: (1) sample no. 1; (2) sample no. 2, and (3) sample no. 3. [148].

The rest of the parameters necessary for calculating the dependence $I = F(D)$ can be found by using the known total concentration of nitrogen atoms in the sample (N_{MAX}).

$$\alpha = I_{MAX}/(N^0_{DAP} + N_0) = I_{MAX}/N$$

$$N^0_{DAP} = (I_0/I_{MAX}) N_{MAX}$$

$$N_0 = N_{MAX}(1 - I_0/I_{MAX})$$

Apparently, the rise in the DPL intensity is due to the increase in the concentration of DAPs via which the radiative recombination occurs. The concentration of DAPs will grow due to the pairing of nitrogen atoms with structural defects formed under irradiation. The dependence $I = F(D)$ will level-off when the concentration of free nitrogen atoms becomes low. Figures 3.9 and 3.10 show the calculated and experimental dependences of the function $I = F(D)$ for all the samples under study. It can be seen in the figures that there is good agreement between the experiment and calculation.

As follows from the table, the initial concentrations of DAPs (N^0_{DAP}) in all the samples under study nearly coincide despite the significant differences in the nitrogen concentration. Apparently, the reason is that the concentration of DAPs is determined by the concentration of structural defects, which is characteristic of the given growth procedure.

There appears a question as to just what structural defect forms a DAP with nitrogen. According to [76,107,152-153], one of the main structural defects whose concentration increases upon irradiation is the carbon vacancy. It can be assumed that the DPL is associated just with nitrogen -V_C pairs, but an acceptor is necessary for acting as a partner of the N donors in the donor-acceptor recombination. V - C is actually a deep acceptor because it represents the negatively charged state in n-SiC, but this issue invites further studies.

A conclusion can also be made from the table that the maximum DPL intensity (i.e., that in saturation) is well correlated with the increase in the concentration of nitrogen atoms. As the nitrogen concentration (N_{MAX}) was raised by an order of magnitude, the I_{MAX} / I_0 ratio increased from 2 to 26.

It can also be noted that the values of γ coincide for p-type samples, but are ~ 400 times smaller for the n-type sample. It is known that the generation rate of Frenkel pairs (n_{FP}) is independent of the conduction type of a SiC sample [154]. Consequently, the main difference between values of the parameter γ is due to the difference either between the constant of interaction of nitrogen atoms with structural defects (β), or between the lifetimes of radiation defects (τ) in n- and p-type SiC. The parameter β is largely determined by the charge states of a nitrogen atom and a structural defect. Because the structural defect mostly acquires a charge corresponding to the sign of majority carriers in a semiconductor, it will be different in p- and n-type materials [155]. If a carbon vacancy is the structural defect, then the parameter τ is determined by direct recombination of this vacancy and an interstitial carbon atom (genetically related) generated together with this vacancy.

Their recombination is also largely due to the charge state of the interacting components. It is quite acceptable that the time τ in the n-type is a hundred of times shorter than that in the p-type. This conclusion agrees with the fact that the carrier removal rate under irradiation with electrons in p-SiC exceeds that in n-SiC [156].

An analysis of experimental results demonstrated that the dependence of the intensity of "defect-related PL" on the irradiation dose is well accounted for on the assumption that the given PL is due to the DAP recombination via nitrogen--radiation defect pairs. The method we suggest can also be used to analyze the radiation-stimulated photoluminescence in other semiconductors.

3.5.2 Restorian of SiC characteristics upon annealing

It is not quite clear whether the irradiated and annealed silicon carbide can be considered equivalent to that in the initial state. This issue may also be of practical importance because the temperature at which radiation defects in silicon carbide are eliminated by annealing is lower than its sublimation point. That is the annealing can restore the initial characteristics of a SiC device without its thermal disintegration. It is of interest whether all characteristics of the material are restored or there are differences.

In [157], the properties of SiC annealed upon irradiation were compared with the properties of the starting material. CVD-grown 4H-SiC samples with uncompensated donor concentration (Nd - Na) $\approx 2 \times 10^{15}$ - 2×10^{16} cm^{-3} were irradiated with 0.9 MeV electrons. The samples were annealed in a vacuum at temperatures of up to 1400°C during 30 min.

The uncompensated donor concentrations (Nd - Na) in the starting and irradiated samples were determined from current-voltage (C-U) characteristics at room temperature. To perform these measurements, Schottky diodes 2 mm in diameter were formed on the surface of an epitaxial layer by deposition of an Au film.

Prior to irradiation, the PL spectrum had a broad band peaked at hv ≈ 2.2 eV, which is commonly related to the radiative recombination via acceptor levels of boron (D-center). Apparently, its presence was due to the uncontrolled doping of the starting epitaxial layers with boron. Only the DPL was resent in the PL spectrum after the irradiation.

The samples under study were at once compensated with a dose of 1×10^{16} cm^{-2} and then annealed. In all, three cycles of this kind were performed. The results obtained are presented in Table 3.6 and Fig. 3.12.

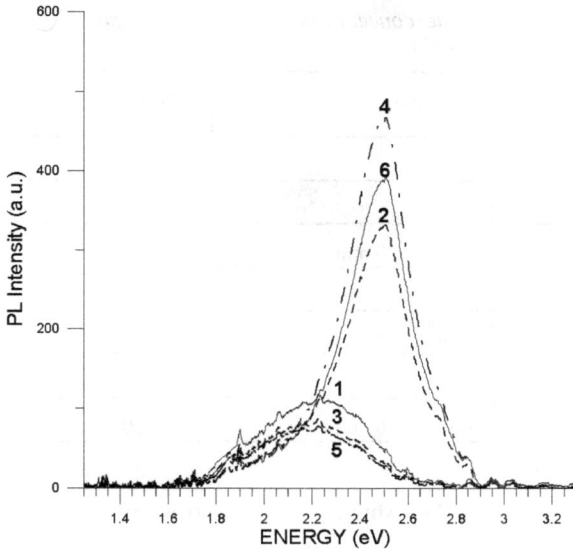

Fig. 3.12. PL spectra: (1) starting sample, (2) sample upon 1st irradiation, (3) that upon 1st annealing, (4) upon 2nd irradiation, (5) upon 2nd annealing, (6) upon 3rd irradiation, and (7) upon 3rd annealing.

It can be seen in the table and figure that the parameters of the samples are nearly completely restored upon annealing. The DPL line disappears from the PL spectrum, and the Nd - Na concentration returns to the initial value to within ~10%. Thus, it can be stated to a first approximation that the irradiation of SiC with 0.9 MeV electrons is reversible. It is not improbable, however, that the irradiation and subsequent annealing result in that some defect complexes are disintegrated and formed, and the change in the concentration of these centers has no effect on the PL spectrum and the experimentally measured value of Nd - Na. The occurrence or absence of changes of this kind in the crystal structure of SiC should be studied by ESR methods.

Table 3,6 Dependence of the concentration Nd-Na of the samples under study upon irradiation and annealing [157].

Doses	Annealing	Nd-Na concentration, cm^{-3}
0		2.6×10^{15}
2×10^{16}		compensated
	1400°C, 30 min	2.43×10^{15}
2×10^{16}		compensated
	1400°C, 30 min	2.49×10^{15}
2×10^{16}		compensated
	1400°C, 30 min	2.4×10^{15}

3.5.3   Point defects in SiC for single-photon spectroscopy

Another kind of PL, associated with structural defects in the crystal lattice of SiC, was considered in [158-164]. It was shown previously that single defects in semiconductors show high promise for application in quantum metrology, quantum calculations, and effective nonclassical sources of light. NV centers (carbon vacancy+ nitrogen atom) in the diamond lattice have been studied in sufficient detail. As regards its properties, a defect of this kind can be considered a single atom weakly bound to the embedding semiconductor matrix. The electron spins of an individual center are easily manipulated by light and by magnetic, electric, and microwave fields, which makes it possible to record quantum information (qubits) on the nuclear spin of the center. This manipulation is even possible at room temperature; the center has long (up to several milliseconds) storage duration of the induced spin polarization. At present, the NV center can be regarded as a base logical element of the future quantum processor required for creating a quantum computer, communication lines with a quantum safety protocol, and other applications of spintronics.

Later, it was demonstrated that the properties similar to those of the NV center in diamond are exhibited by silicon vacancies (V_{Si}) in the SiC lattice. The V_{Si} defect is constituted by a silicon vacancy and a neighboring carbon vacancy, both arranged along the C axis of the SiC crystal. This complex possesses a number of spin-optical properties that are rather promising for application in quantum technologies, which makes it possible to regard this complex as an alternative to the known NV defect in diamond. The centers were introduced via neutron irradiation of SiC crystals (dose $\sim 10^{16}$ cm^{-2}), followed by annealing at ~ 2000 °C.

Owing to the existence of a large number of silicon carbide polytypes and nonequivalent positions in the silicon carbide lattice, the properties of the centers are much more diverse than those of the NV center in diamond. As recently demonstrated in experiments on optically detected magnetic resonance (ODMR), V_{Si} centers have the ground spin state S = 3/2. This state can be optically pumped, which leads to a dependence of the PL intensity on the spin orientation. This property can be used as a basis for various sensor applications, with the shift of resonance line measured as a function of external magnetic fields and temperature.

References

[1]   N. D. Sorokin, Yu. M. Tairov, V. F. Tsvetkov, and M. A. Chernov, "Investigation of the crystal-chemical properties of the silicon carbide polytypes", Kristallografiya 28, 910-914 (1983).

[2]   Vodakov Yu A, Lomakina G A and Mokhov E N, Non-stoichiometry and polytypism of silicon carbide Sov.Phys. Solid State 24 780 (1982).

[3]   I. M. Pavlov, M. I. Iglitsin, M. G. Kosaganov, and V. N. Solomatin, "Centers with spin 1 in silicon carbide, irradiated by neutrons and a-particles", Sov. Phys. Semicond. 9, 1320-1326 (1975).

[4]   A. I. Veinger, A. A. Lepeneva, G. A. Lomakina. E. N. Mokhov, and V. I.Sokolov, Sov., "Annealing of the radiation defects in n-SiC(6H), irradiated by neutrons", Phys. Semicond. 18, 1932-1937 (1984).

[5]   A. A. Lebedev and N.A. Sobolev "Capacitance spectroscopy of deep centers in SiC" Material Science Forum 258–263, 715–720 (1997).

[6]   M.M.Anikin, A.A.Lebedev, A.L.Syrkin, and A.V.Suvorov, Investigation of deep levels in SiC by capacitance spectroscopy, Sov.Phys.Semicond, 19, 69-71,(1985).

[7]   Lamakina G A, Vodakov Yu A, Mokhov E N, Oding V G and Kholuyanov G F Comparative studies of the three polytypes of silicon carbide, Sov. Phys. Solid State 12 2356, (1970).

[8]   M. M. Anikin, A. A. Lebedev, A. A. Lebedev, A. L. Syrkin, and A. V. Suvorov, Capacitance spectroscopy of pn junction made of epitaxial 4H-SiC, doped by implantation of Al ions, Sov. Phys. Semicond. 20, 1357-1359 (1986).

[9]   N. I. Kuznetsov and J. A. Edmond, deep level s influence on current relation in 6H-SiC diodes", Sov. Phys. Semicond. 31, 1481-1485 (1997).

[10]   Matsuura H, Komeda M, Kagamihara S, Iwata H, Ishihara R, Hatakeyama T, Watanabe T, Kojima K, Shinohe T and Arai K Dependence of acceptor levels and hole mobility on acceptor density and temperature in Al-doped p-type 4H-SiC epilayers 2004 J. Appl. Phys. 96 2708–15

[11]   A A Lebedev, P L Abramov, E V Bogdanova, S P Lebedev, D K Nel'son, G A Oganesyan, A S Tregubova and R Yakimova, Highly doped p-type 3C–SiC on 6H–SiC substrates, Semicond. Sci. Technol. 23 (2008) 075004 (4pp). http://dx.doi.org/10.1088/0268-1242/23/7/075004

[12]   A. N. Andreev, M. M. Anikin, A. A. Lebedev, N. K. Poletaev, A. M. Strel'chuk, A. L. Syrkin, and V. E. Chelnokov, Relationship between "defect" luminescence in 6H SiC and deep-level centres, Semiconductors 28, 430-435 (1994).

[13]   A. O. Konstantinov, About nature of the point defects, arising during diffusion of the acceptor impurity in silicon carbide, Sov.Phys. Semicond. 26, 151-158 (1992).

[14]   Y.Gao, S.I.Soloviev and T.S.Sudarshan, "Investigation of boron diffusion in 6H-SiC" Appl.Phys.Lett. 83 905–907 (2003). http://dx.doi.org/10.1063/1.1598622

[15]   M. M. Anikin, A. A. Lebedev, N. K. Poletaev, A. M. Strel'chuk, A. L.Syrkin, and V. E. Chelnokov, " Deep centers and blue-green electroluminescence in 4H-SiC" Inst. Phys. Conf. Ser. 137, Chap. 6, 605–607, (1994).

[16]   M. S. Janson, M. K. Linnarson, A. Hallen, B. G. Svensson, N. Nordel and H. Bleichner, "Transient enhanced diffusion of implanted boron in 4H-silicon carbide", Appl.Phys.Lett. 76, 1434–1436 (2000). http://dx.doi.org/10.1063/1.126055

[17]   M. Laube, G. Pensl and H. Itoh, "Suppressed diffusion of implanted boron in 4H-SiC", Appl.Phys.Lett. 74, 2292–2294 (1999). http://dx.doi.org/10.1063/1.123828

[18]   H. Bracht, N. A. Stolwijk, M. Laube, G. Pensl, "Diffusion of boron in silicon carbide: Evidence for the kick-out mechanism" Appl.Phys.Lett. 77, 3188–3190 (2000). http://dx.doi.org/10.1063/1.1325390

[19]   M. S. Mazzola, S. E. Saddow, P. G. Neudeck, V. K. Lakdawala, and S. We," Observation of the D-centre in 6H-SiC pn diodes grown by chemical vapor deposition" Appl. Phys. Lett. 64. 2730-2733(1994). http://dx.doi.org/10.1063/1.111457

[20] M. M. Anikin, N. I. Kuznetsov, A. A. Lebedev, N. K. Poletaev, A. M. Strel'chuk, A. L. Syrkin and V. E. Chelnokov, " Shift of the electroluminescence peak in 6H-SiC based diodes with the forward current density", Semiconductors 28, 270-273 (1994).

[21] J. P. Doyle, M. 0. Adoelfotoh, B. G. Svensson, A. Schoner, and N. Nordel, Characterization of electrically active deep level defects in 4H and 6H SiC, Diamond Relat. Mater. 6, 1388–1391 (1997). http://dx.doi.org/10.1016/S0925-9635(97)00102-7

[22] T. Dalibor, G. Pensl, H. Matsunami, T. Kimoto, W. J. Choyke, A. Schoner, and N. Nordel, Deep defect Centers in Silicon Carbide Monitored with deep Level Transient Spectroscopy Phys. Status Solidi A 162, 199–232 (1997). http://dx.doi.org/10.1002/1521-396X(199707)162:1<199::AID-PSSA199>3.0.CO;2-0

[23] A.V.Bolotnikov, P.G.Muzykov, T.S.Sudarshan, Investigation of two-branch boron diffusion from vapor phase in n-type 4H-SiC, Appl.Phys.Lett 93 (2008) 052101 http://dx.doi.org/10.1063/1.2968306

[24] A.A.Lebedev, Deep levels defects in silicon carbide, Intern. J. of High Speed Electronics and Systems. V 16, N 3 (2006) 779 -823. http://dx.doi.org/10.1142/S0129156406004028

[25] P. G. Baranov and E. N. Mokhov, Electron paramagnetic resonance of deep boron in SiC, Inst. Phys. Conf. Ser. 142, 293–296 (1996).

[26] A. van Duijn-Arnold, J. Moi, R. Verberk, J. Schmidt, E. N. Mokhov, and P. G. Baranov, Spatial distribution of the electronic wave function of the shallow boron acceptor in 4H- and 6H-SiC, Phys.Rew. B., 60, 15829–15847 (1999). http://dx.doi.org/10.1103/PhysRevB.60.15829

[27] S. Jang, T. Kimoto, and H. Matsunami, Deep levels in 6H-SiC wafers and step-controlled epitaxial layers, Appl. Phys. Lett. 65, 581–583 (1994). http://dx.doi.org/10.1063/1.112302

[28] V. S. Balandovich and G. N. Violina, "Photocapacitance effect in silicon carbide, doped by boron", Sov. Phys. Semicond. 15, 959-961 (1981).

[29] A.A. Lebedev and N.K.Poletaev, High temperature "boron" electroluminescence in 4H-SiC and deep centers, Inat.Phys.Conf Ser 142, chapter 2, 349-352, (1996)

[30] A.A.Lebedev, Deep level centers in silicon carbide: A review, Semiconductors 33, 107-130 (1999). http://dx.doi.org/10.1134/1.1187657

[31]   E. N. Kolabukova, S. N. Lukin, E. N. Mokhov, J. Reinke, S. Greulich-Weber, and J.-M. Spaeth, "New deep acceptor at Ev+0,8 eV in 6H silicon carbide", Inst. Phys. Conf. Ser. 142, 333–335 (1996).

[32]   S.W.Huh, H.J.Chung, S.Nigam, A.Y.Polyakov, M.Skowronski, E.R.Glassek, W.E. Carlos, B.V.Shanabrook, M.A.Fanton, N.B.Smirnov, Residual impurities and native defects in 6H-SiC bulk crystals grown by halide chemical-vapor deposition, J.Appl.Phys 99, (2006) 013508 http://dx.doi.org/10.1063/1.2150593

[33]   G.Alferi and T.Kimoto, High – temperature annealing behavior of deep levels in 1 MeV electron irradiated 6H-SiC, Appl.Phys.Lett. 93, 032108, (2008). http://dx.doi.org/10.1063/1.2964184

[34]   K.Dano, T.Kimoto, H.Mtsunami, Midgap levels in both n- and p-type 4H-SiC epilayers investigated by deep level transient spectroscopy. Appl.Phys.Lett 86 (2005) 122104 http://dx.doi.org/10.1063/1.1886904

[35]   M. M. Anikin, A. S. Zubrilov, A. A. Lebedev, A. M. Strel'chuk, and A. E. Cherenkov, Recombination processes in 6H-SiC pn structures and the influence of deep centers, Sov. Phys. Semicond. 25, 289-293 (1991).

[36]   C.G.Hemmigson, N.T.Son, O.Kordina and E.Janzen, Metastable defects in 6H-SiC: experiments and modeling, J.Appl.Phys 91, 1324–1330 (2002). http://dx.doi.org/10.1063/1.1427401

[37]   H. Zhang, G. Pensl, A. Domer, and S. Leibenzeder, Deep centers in n-type 6H-SiC, Ext. Abstr. Electrochem. Soc. Mtg., 699–700 (1989).

[38]   J.P. Doyle, M.O. Adoelfotoh, B.G. Svensson, A. Schoner, N. Nordel, Characterization of electrically active deep level defects in 4H and 6H SiC, Diamond and Related Mater., 1388 (1997).

[39]   A. Kawasuso, F. Redmann, R. Krause-Rehberg, P. Sperr, Th. Frank, M. Weidner, G. Pensl and H. Itoh, Vacancies and deep levels in electron-irradiated 6H-SiC epilayers studied by positron annihilation and deep level transient spectroscopy, J.Appl.Phys, 90, 3377–3382 (2001). http://dx.doi.org/10.1063/1.1402144

[40]   C. Rubicki, Deep level defects in alpha participle irradiated 6H silicon carbide, J. Appl. Phys. 78, 2996–3000 (1995). http://dx.doi.org/10.1063/1.360048

[41]   A. I. Veinger, V. A. Il'in, Yu. M. Tairov and V. F. Tsvetkov, Investigation parameters of the paramagnetic deep centers of the vacancies nature in 6H SiC,Sov.Phys.Semicond. 15, 902-908 (1981).

[42]   T. Dalibor, G. Pensl, N. Nordel, A. Schoner, and W. J. Choyke, Ground States of
       the Ionized Iso-electronic Ti Acceptor in SiC, Material Science Forum 264–268,
       537–540 (1998). http://dx.doi.org/10.4028/www.scientific.net/MSF.264-268.537

[43]   A. A. Lebedev, D. V. Davydov, N. S. Savkina, A. S. Tregubova, M. P. Scheglov,
       R. Yakimova, M. Suvajarvi and E. Janzen, Structural Defects and Deep-Level
       Centers in 4H-SiC epilayers grown by sublimation epitaxy in vacuum
       Semiconductors, 34, 1183-1136 (2000). http://dx.doi.org/10.1134/1.1317570

[44]   T. Dalibor, G. Pensl, T. Kimoto, H. Matsunami, S. Shidhara, R. P. Devaty, and W.
       J. Choyke, Radiation-induced defect centers in 4H silicon carbide, Diamond Relat.
       Mater. 6, 1333–1337 (1997). http://dx.doi.org/10.1016/S0925-9635(97)00108-8

[45]   I.Pintilie, L.Pintilie, K.Irmscher and B.Thomas, Formation of the Z1,2 deep-level
       defects in 4H-SiC epitaxial layers: Evidence for nitrogen participation"
       Appl.Phys.Lett 81, 4841–4843 (2002) http://dx.doi.org/10.1063/1.1529314

[45]   K.Danno and T.Kimoto, Investigation of deep levels in n-type 4H-SiC epilayers
       irradiated with low-energy electrons, J.Appl.Phys 100, 113728 (2006)
       http://dx.doi.org/10.1063/1.2401658

[46]   M.Asghar, I.Hussain, H.S.Noor, F.Iqbal, Q.Wahab and A.S.Bhatti, Properties of
       dominant electron trap center in n-type SiC epilayers by means of deep level
       transient spectroscopy, J.Appl.Phys 101, 073706 (2007)
       http://dx.doi.org/10.1063/1.2715534

[47]   L.S.Lovilie and B.G.Svensson, Enhanced anneling of implantation-induced defects
       in 4H-SiC by thermal oxidation, Appl.Phys Lett. 98, 052108 (2011).
       http://dx.doi.org/10.1063/1.3531755

[48]   H.M.Ayedh, R.Nipoti, A.Hallen and B.G.Svensson, Isothermal Treatment Effects
       on the Carbon Vacancy in 4H Silicon Carbide, Mat.Sci Forum, 821-823, (2015)
       351-354. http://dx.doi.org/10.4028/www.scientific.net/MSF.821-823.351

[49]   K.Kawahara, X.T. Trinh, N.T. Son, E.Janzen, J.Suda and T.Kimoto, Quantitative
       comparison between Z1/2 center and carbon vacancy in 4H-SiC, J.Appl.Phys 115,
       143705 (2014). http://dx.doi.org/10.1063/1.4871076

[50]   H. M. Ayedh, V. Bobil, R. Nipoti. A. Hallen and B. G. Svensson, Formation and
       Annihilation of Carbon Vacancies in 4H-SiC, Mat.Sci Forum, 858, (2016) 331-
       336. http://dx.doi.org/10.4028/www.scientific.net/MSF.858.331

[51]   S.A.Reshanov and G.Pensl, Comparasion of Electrically and Optically Determined Minority carrier Lifetimes in 6H-SiC, Mater. Science Forum 483–485, 417 (2005). http://dx.doi.org/10.4028/www.scientific.net/MSF.483-485.417

[52]   J. R. Jenny, D. P.Malta, V. F. Tsvetkov, M. K. Das, H. D. Hobgood, C. Y. Carter, R.J. Kumar, J. M. Borrego, R. J Gutmann, R. Aavikko, Effects of annealing on carrier lifetime in 4H-SiC, JAP 100 (2006) 113710. http://dx.doi.org/10.1063/1.2372311

[53]   R.J.Kumar, J.M.Borrego, R.J.Gutmann, J.R.Jenny, D.P.Malta, H.D.Hobgood, C.Y.Carter, Microwave photoconductivity decay characterization of high-purity 4H-SiC substrate, JAP 102 (2007) 013704. http://dx.doi.org/10.1063/1.2751086

[54]   L.Storasta, H.Tsuchida, T. Miyazawa, T. Ohsima, Enchanced annealing of Z1/2 defect in 4H-SiC epilayers, JAP, 103 (2008) 013705.

[55]   A.O.Evwaraye, S.R.Smith, W.C.Mitchel, G.C.Farlow, Electric field enhancement of electron emission rates from Z1/2 centers in 4H-SiC, JAP, 106 (2009) 063702.

[56]   K.Danno, D.Nakamura, T.Kimoto, Investigation of carrier lifetime in 4H-SiC epilayers and lifetime control by electron irradiation, APL, 90 (2007) 202109. http://dx.doi.org/10.1063/1.2740580

[57]   H. M. Ayedh, A. Hallen, and B. G. Svensson, Elimination of carbon vacancies in 4H-SiC epi-layers by near-surface implantation: Influence of the ion species" J.Appl.Phys 118, 175701 (2015). http://dx.doi.org/10.1063/1.4934947

[58]   I.D. Booker, E. Janzen, N.T.Son, J.Hassan, P.Stenberg, and E.O. Sveinbjornsson, Donor and double-donor transitions of the carbon vacancy related EH6/7 deep level in 4H-SiC, J.Appl.Phys 119, 235703 (2016). http://dx.doi.org/10.1063/1.4954006

[59]   J.Weber, S.Beljakova, H.B.Weber, G.Pensl, B.Zippelius, T.Kimoto, M.Kriger, Determination on the Electrical capture Process of the EH6-Center in n-type 4H-SiC, Mater. Science Forum 740-742, 377-380 (2013). http://dx.doi.org/10.4028/www.scientific.net/MSF.740-742.377

[60]   S.Sasaki, K.Kawahara, G.Feng, G.Alferi, and T.Kimoto, Major deep levels with the same microstructures observed in n-type 4H-SiC and 6H-SiC, J.Appl.Phys 109, 013705 (2011). http://dx.doi.org/10.1063/1.3528124

[61]   J. M. Langer and H. Heinrich, "Deep-level impurities: A possible guide to prediction of band-edge discontinuities in semiconductor heterojunctions, Phys. Rev. Lett. 55, 1414–1417 (1985). http://dx.doi.org/10.1103/PhysRevLett.55.1414

[62]   P. Zhou, M. G. Spencer, G. L. Harris, and K. Fekade, Observation of deep levels
        in cubic silicon carbide, Appl. Phys. Lett. 50, 1384–1385 (1987).
        http://dx.doi.org/10.1063/1.97864

[63]   K. Zakentes, M. Kayiambaki, and G. Constandnidis, Electron traps in ?-SiC grown
        by chemical vapor deposition on silicon (100) substrates, Appl. Phys. 66, 3015–
        3017 (1995). http://dx.doi.org/10.1063/1.114262

[64]   D. V. Davydov, A. A. Lebedev, A. S. Tregubova, V. V. Kozlovskii, A. N.
        Kuznetsov, and E. V. Bogdanova, Investigation of 3C-SiC Epitaxial Layers Grown
        by Sublimation Etpitaxy, Mater.Science Forum 338–342, 221–224 (2000).
        http://dx.doi.org/10.4028/www.scientific.net/MSF.338-342.221

[65]   G.Alferi, H.Nagasawa, T.Kimoto, Thermal stability of deep levels between room
        temperature and 1500 0C in as-grown 3C-SiC, JAP, 106 (2009) 073721.

[66]   V. S. Balandovich and G. N. Violina, An Investigation of radiation defects in
        silicon carbide irradiated with fast electrons, Cryst. Lattice Defects Amorphous
        Mater. 13, 189–193 (1987).

[67]   H. Zhang, G. Pensl, P. Glasow, S. Leibenzeder. Ext. Abstracts Electrochem. Soc.
        Mtg., 714 (1989).

[68]   X.D. Chen, C.L.Yang, M.Gong, W.K.Ge, S.Fung, C.D.Beling, J.N.Wang,
        M.K.Lui, and C.C.Ling, Low Energy Electron Irradiation Induced Deep Leve
        Defects in 6H-SiC: The Implication for the Microstructure of the Deep Levels
        E1/E2, Phys Rew Lett 92 N12, 125504 (2004).
        http://dx.doi.org/10.1103/PhysRevLett.92.125504

[69]   J.M.Luo, Z.Q. Zhong, M.Gong, S. Fung and C.C.Ling, Isochronal annealing study
        of low energy electron irradiated Al-doped p-type 6H silicon carbide with deep
        level spectroscopy, J.Appl.Phys 103, 063711 (2009).
        http://dx.doi.org/10.1063/1.3087757

[70]   M. Gong, S. Fung, C. D. Beiling, Zhipu You, A deep level transient spectroscopy
        study of electron irradiation induced deep levels in p-type 6H-SiC, J.Appl.Phys 85,
        7120 –7122 (1999). http://dx.doi.org/10.1063/1.370520

[71]   C. Hemmingson, N. T. Son, 0. Kordina, E. Janzen. J. L. Lindstrom, S. Savarge,
        and N. Nordel, Capacitance transient studies of electron irradiated 4H-SiC, Mater.
        Sci. Eng. B 46. 336–339 (1997). http://dx.doi.org/10.1016/S0921-5107(96)01999-
        X

[72]   C. Hemmingson, N. T. Son, 0. Kordina, E. Janzen. J. P. Bergman, J. L. Lindstrom, S. Savarge, and N. Nordel, Deep level defects in electron-irradiated 4H SiC epitaxial layers, J.Appl.Phys. 81, 6155–6159 (1997). http://dx.doi.org/10.1063/1.364397

[73]   M.L.David, G.Alferi, E.M.Monakhov, A.Hallen, C.Blanchard, B.G.Svensson, J.F.Barbot, Electrically active defects in irradiated 4H-SiC, J.Appl.Phys 95 4728–4733 (2004). http://dx.doi.org/10.1063/1.1689731

[74]   I.Pintilie, U.Grossner, B.G.Svensson, K.Irmscher, and B.Thomas, Influence of growth conditions on irradiation induced defects in low doped 4H-SiC epitaxial layers, Appl.Phys.Lett. 90, 062113 (2007). http://dx.doi.org/10.1063/1.2472173

[75]   A. Castaldini, A. Cavallini, L. Rigutti, and F. Nava, Low temperature annealing of electron irradiation induced defects in 4H-SiC, Appl.Phys.Lett. 85, 3780 (2004). http://dx.doi.org/10.1063/1.1810627

[76]   Z.Zolnai, N.T.Son, C.Hallin, and E.Janzen, Annealing behaviour of the carbon vacancy in electron-irradiated 4H-SiC, J.Appl.Phys. 96, 2406 (2004). http://dx.doi.org/10.1063/1.1771472

[77]   K.Danno and T. Kimoto, Deep level transient spectroscopy on as-grown and electron-irradiated p-type 4H-SiC epilayers, J.Appl.Phys. 101, 103704 (2007). http://dx.doi.org/10.1063/1.2730569

[78]   H.Matsuura, N.Minohara, and T. Ohshima, Mechanisms of unexpected reduction in hole concentration in Al-doped 4H-SiC by 200 keV electron irradiation, J.Appl.Phys. 104, 043702 (2008). http://dx.doi.org/10.1063/1.2969788

[79]   K. Yoshihara, M. Kato, M. Ichimura, T. Hatayama and T. Ohshima, Deep levels in p-type 4H-SiC induced by low-energy electron irradiation, Mater.Sci.Forum 70-742, (2013) 373-376.

[80]   H Matsuura, K. Aso, S.Kgamhila, H.Iwata, T.Ishida and K.Nishikawa, Decrease in Al acceptor density in Al-doped 4H-SiC by irradiation with 4,6 MeV electrons, Appl.Phys.Lett., 83 N 24,(2003) 4981-4983. http://dx.doi.org/10.1063/1.1634381

[81]   N.T.Son, B.Magnusson and E.Janzen, Photoexitation-electron-paramagnetic-resonance studies of the carbon vacancy in 4H-SiC, Appl.Phys.Lett. 81 3945–3947 (2002). http://dx.doi.org/10.1063/1.1522822

[82]   A. I. Veinger, A. A. Lepeneva, G. A. Lomakina. E. N. Mokhov, and V. I.Sokolov, Sov., "Annealing of the radiation defects in n-SiC(6H), irradiated by neutrons", Phys. Semicond. 18, 1932-1937 (1984).

[83] R. N. Kyutt, A. A. Lepeneva, G. A. Lomakina, E. N. Mokhov, A. S. Tregubova, M. P. Scheglov and G. F. Uldashev, "Formation of the defects during annealing of the neutron irradiated silicon carbide", Sov.Phys. Solid State 30, 2606-2610 (1988).

[84] A. A. Lepeneva, E. N. Mokhov, V. G. Oding and A. S. Tregubova, Silicon carbide, irradiated by high dozes of neutrons", Solid State Physics 33, 2217-2221 (1991).

[86] M. Okada, T. Kimura, T. Nakata, M. Watanbe, S. Kanazava, I. Kanno, K. Kamitani, K. Atobe, and M. N. Nakagawa, "Radiation-induced defects in neutron-irradiated 4H- and 6H-SiC single crystals", Inst. Phys. Conf. Ser. 142,469–472 (1996).

[87] H. Matsunami and T. Kimoto, "Step controlled epitaxial growth of SiC: high quality homoepitaxy", Mater. Sci. Eng., R. 20, 125–166 (1997).

[88] V.Nagesh,J.W.Farmer,R,F,Davis and H,S,Kong, "Defects in neutron irradiated SiC", Appl.Phys.Lett. 50, 1138–1140 (1987). http://dx.doi.org/10.1063/1.97941

[89] Th.Lingner, S.Greulich-Weber, J.-M. Spaeth, U.Gerstmann, E.Rauls, Z.Hajnal, Th.Frauenheim, and H.Overhof, "Structure of silicon vacancy in 6H-SiC after annealing identified as the carbon vacancy-carbon antisite pair", Phys.Rev. B 64, 245212 (2001). http://dx.doi.org/10.1103/PhysRevB.64.245212

[90] I.V.Ilyin, M.V.Muzafarova, E.N.Mokhov, S.G.Konnikov and P.G.Baranov, "Multi-defect Clusters in Neutron Irradiated Silicon Carbide: Electron Paramagnetic Resonance Study" Mater. Science Forum 483–485, 489 (2005).

[91] I. V. Ilyin, E. N. Mokhov, P. G. Baranov, Paramagnetic defects in silicon carbide crystals irradiated with gamma-ray quanta, Semiconductors, 35, pp 1347–1354, (2001). http://dx.doi.org/10.1134/1.1427968

[92] P.G.Baranov, B.Ya.Ber, I.V.Ilyin, A.N.Ionov, M.V.Muzafarova, M.A.Kaliteevskiaa, P.S.Kop'ev, A.K. Kaliteevskii, O.N. Godisov, and I.M.Lazebnik, "Peculiarities of neutron-tranmution phosphoros doping of 30Si enriched SiC crystals: Electron paramagnetic resonsnce study" J.Appl.Phys. 102, 063713, (2007). http://dx.doi.org/10.1063/1.2783884

[93] G.Alferi, E.V.Monakhov, B.G.Svensson, and A.Hallen, Defect energy in hydrogen-implanted and electron-irradiated n-type 4H silicon carbide, J.Appl.Phys. 98, 113524 (2005). http://dx.doi.org/10.1063/1.2139831

[94]   A. A. Lebedev, D. V. Davydov, A. M. Strel'chuk, V. V. Kozlovskii, A. N. Kuznetsov, and E. V. Bogdanova, "Deep centers appearing in 6H and 4H SiC after proton irradiation", Mat.Science 338-342, 973-976 (2000).

[95]   A. M. Strel'chuk, V. V. Kozlovskii, N. S. Savkina, M. G. Rastegaeva, A. N. Andreev. Influence of proton irradiation on recombination current in 6H-SiC pn structures, Mat.Science & Eng. 61-62, 441-445 (1999). http://dx.doi.org/10.1016/S0921-5107(98)00551-0

[96]   A. A. Lebedev, A. M. Strel'chuk, V. V. Kozlovskii, N. S. Savkina, D. V. Davydov, and V. V. Soloviev. Studies of the effect of proton irradiation on 6H-SiC pn junction properties, Mat.Science & Eng. 61-62, 450-453 (1999). http://dx.doi.org/10.1016/S0921-5107(98)00553-4

[97]   M.F.Barthe, P.Desgardin, L.Henry, C.Corbel, D.T.Britton, G.Kogel, S.Sperr, W.Trifftshauser, P.Vicente, L. di Ciocco, Vacancy Defects in As-Polished and in High-Fluence H+ - implanted 6H-SiC Detected by Slow Positron Annihilation Spectroscopy, Material Science Forum, 389-393, 493 (2002). http://dx.doi.org/10.4028/www.scientific.net/MSF.389-393.493

[98]   D.T.Britton, M.F.Barthe, C.Corbel, A.Hempel, L.Henry, P.Desgardin, W.Bauer-Kugelman, G.Kogel, S.Sperr, W.Triftshauser. Evidence for negatively charged vacancy defects in 6H-SiC after low-energy proton implantation Appl.Phys.Lett. 78, 1234 (2001). http://dx.doi.org/10.1063/1.1350961

[99]   H. J. von Bardeleben, J. L. Cantin, I. Vickridge and G. Battisting. Proton-implantation-induces defects in n-type 6H- and 4H-SiC. An electron paramagnetic resonance study. Phys.Rew. 62 10126-10134 (2000). http://dx.doi.org/10.1103/PhysRevB.62.10126

[100] L.Henry, M.-F. Barthe, C.Corbel, P. Desgardin, G.Blondiaux, S. Appianen, and L.Liszkay, Silicon vacancy-type defects in as-received and 12-MeV proton-irradiated 6H-SiC studied by positron annihihilation spectroscopy, Phys.Rev B 67, 115210 (2003). http://dx.doi.org/10.1103/PhysRevB.67.115210

[101] D.V.Davydov, A.A.Lebedev, V.V.Kozlovski, N.S.Savkina, A.M.Strel'chuk. "DLTS study of defects in 6H- and 4H-SiC created by proton irradiation " Physica B 308-310, 641 (2001). http://dx.doi.org/10.1016/S0921-4526(01)00775-X

[102] A.O.Konstantinov, V.N.Kuzmin, L.S.Lebedev, D.P.Litvin, A.G.Ostroumov, V.I.Sankin, V.V.Semenov, Effect of proton radiation on the electrical properties of silicon carbideSov.Tech. Phys, 54, 1622 (1984).

[103] А.О.Константинов, Н.С.Константинова, О.И.Коньков, Е.И.Теруков, Passivation crystalline silicon carbide in the hydrogen plasma, Sov., Phys., Semicond., 28, 342 (1994).

[104] P.?. Ivanov, O.I.Kon'kov, V.N.Panteleev, T.P.Samsonova, Influence of plasma treatment of silicon carbide on the surface characteristics of FETs concealed p-n-gate, Semiconductors, 31, 1404 (1997).

[105] N. Atchtziger, J. Grillenberger, W. Witthuhn, M. K. Linnarsson, J. Janson, B. G. Svensson . Hydrogen passivation of silicon carbide by low-energy ion implantation Appl.Phys.Lett. 73, 945 (1998) http://dx.doi.org/10.1063/1.122047

[106] R.K.Nadella and M.A.Capano. High-resistance layers in n-type 4H-silicon carbide by hydrogen ion implantationAppl.Phys.Lett 70, 886 (1997).

[107] A. A. Lebedev, A. I. Veinger, D. V. Davydov, A. M. Strel'chuk, V. V. Kozlovskii, N. S. Savkina " Doping of n-type 6H-SiC and 4H-SiC with defects created with a proton beam", J.Appl.Phys. 88, 6265–6271 (2000). http://dx.doi.org/10.1063/1.1309055

[108] G.Alferi, A.Mihaila, H.M.Ayedh, B.G.Svensson, P.Nazdra, P.Godignon, J.Milan, S.Kicin, Deep Level Characterization of 5 MeV proton irradiated SiC PiN diodes, Mat.Sci.Forum, V 858, 308-311 (2016). http://dx.doi.org/10.4028/www.scientific.net/MSF.858.308

[109] K.Kawahara, G.Alfieri and T.Kimoto, Detection and depth analyses of deep levels generated by ion implantation in n- and p-type 4H-SiC, J.Appl.Phys, 106, 013719 (2009). http://dx.doi.org/10.1063/1.3159901

[110] K.Kawahara, J.Suda, G.Pensl, and T.Kimoto, Reduction of deep levels generated by ion implantation into n- and p-type 4H-SiC, J.Appl.Phys, 108, 033706 (2010). http://dx.doi.org/10.1063/1.3456159

[111] T.Sledziewski, G.Ellortt, W. Rosch, H.B.Weber and M.Krieger, Reduction of implantation-induced point defects by germanium ions in n-type 4H-SiC, Mater.Sci.Forum 821-823, 347-350 (2015). http://dx.doi.org/10.4028/www.scientific.net/MSF.821-823.347

[112] T.Sledziewski, S.Beljakova, K. Alassd, P. Kwasnicki, R. Arvinte, S. Jullaguet, M.Zielinki, V. Souliere, G.Ferro, H.B.Weber and M.Krieger, Characterization of Ge - doped homoepitaxial layers grown by chemical vapor deposition, Mater.Sci. Forum V 778- 780, (2014) 261

[113] J.Cotton, G.Gruber, P.Hadley, M.Koch, G.Podegen, T.Aichiger and A.Shuluder, Recombination centers in 4H-SiC investigated by electrically detected magnetic resonsnce and ab initio modeling, J.Appl.Phys, 119, 181507 (2016). http://dx.doi.org/10.1063/1.4948242

[114] J.B.Allance, L.B.Bayu Aji, T.T.Li, L.Shao and S.O. Kucheyev, Damage buildup in Ar-ion-irradiated 3C-SiC at elevated temperatures, J.Appl.Phys, 118, 105706 (2015). http://dx.doi.org/10.1063/1.4930799

[115] A.V. Naumov, V.I.Sankin, The lifetime of the nonequilibrium holes in diodes based on SiC, Sov.Phys.Semicond., 23, 1009 (1989).

[116] V.I.Sankin, R.G.Verenchikova, Ya.A.Vodakov, M.G.Ramm, A.D.Roenkov, "The diffusion length of minority carriers in 6 H and 4H SiC ", Sov.phys.Semicond. 16, 1325 (1983).

[117] V.S.Balandovich, G.N.Violina Photocapacitance effect in silicon carbide doped with boron, Sov.phys.Semicond. 15, 1650 (1981).

[118] M. M. Anikin, A. A. Lebedev, S. N. Pyatko, V. A. Soloviev, and A. M. Strel'chuk, "Minority Carrier Diffusion Length in epitaxially grown SiC(6H) pn diodes" Springer Proc. Phys. 56, 269–273 (1992). http://dx.doi.org/10.1007/978-3-642-84402-7_40

[119] A.M.Strel'chuk, The lifetime and diffusion length of nonequilibrium carriers in SiC p-n structures, Sov.phys.Semicond. 29, 1190(1995).

[120] St G. Muller, D. Hofmann, A. Winnacker, E. N. Mokhov, and Yu. A.Vodakov, "Vanadium as a recombination centre in 6H SiC", Inst. Phys. Conf. Ser. 142, 361–364 (1996).

[121] A.M. Strel'chuk and V. V. Evstropov, "Dominate recombination centre in 6H SiC", Inst. Phys. Conf. Ser. 155, 1009–1012 (1997).

[122] O. Kordina, J. P. Bergman, A. Henry, and E. Janzen, "Long minority carrier lifetime in 6H-SiC grown by chemical vapor deposition", Appl. Phys. Lett. 66, 189–191 (1995). http://dx.doi.org/10.1063/1.113130

[123] J. P. Bergman, E. Janzen, S.G.Sridhara and W.J.Choyke, "Time resolved PL Study of Multi Bound Exitons in 3C SiC", Material Science Forum, 264–268,485–488 (1998). http://dx.doi.org/10.4028/www.scientific.net/MSF.264-268.485

[124] N. V. Daykonova, P. A. Ivanov, V. A. Kozlov, M. E. Levinshtein, J. W. Palmor, S. L. Rumyantsev and R. Singh, "Steady-State and Transient Forward Current-Voltage Characteristics of 4H-Silicon Carbide 5,5 kV Diodes at High and

Superhigh Current density", IEEE transaction on Electron Devioces 46, 2188–2194.

[125] P. A. Ivanov, M. E. Levinshtein, T. T. Mnatsakanov, J. W. Palmour, A. K. Agarwal Power bipolar devices based on silicon carbide, Semiconductors, 39, 861 (2005). http://dx.doi.org/10.1134/1.2010676

[126] B.R.Gossik, "On the Transient Behaviour of Semiconductor Rectifiers" J.Appl.Phys, 27, 905 (1956). http://dx.doi.org/10.1063/1.1722512

[127] S.A.Reshanov and G.Pensl "Comparasion of Electrically and Optically Determined Minority carrier Lifetimes in 6H-SiC" Mater. Science Forum 483–485, 417 (2005). http://dx.doi.org/10.4028/www.scientific.net/MSF.483-485.417

[128] J.R.Jenny, D.P.Malta, V.F.Tsvetkov, M.K.Das, H.D.Hobgood, C.Y.Carter, R.J.Kumar, J.M.Borrego, R.J.Gutmann, R.Aavikko, Effects of annealing on carrier lifetime in 4H-SiC, JAP 100 (2006) 113710. http://dx.doi.org/10.1063/1.2372311

[129] R.J.Kumar, J.M.Borrego, R.J.Gutmann, J.R.Jenny, D.P.Malta, H.D.Hobgood, C.Y.Carter. Microwave photoconductivity decay characterization of high-purity 4H-SiC substrate, JAP 102 (2007) 013704. http://dx.doi.org/10.1063/1.2751086

[130] L.Storasta, H.Tsuchida, T. Miyazawa, T. Ohsima, Enchanced annealing of Z1/2 defect in 4H-SiC epilayers, JAP, 103 (2008) 013705.

[131] A.O.Evwaraye, S.R.Smith, W.C.Mitchel, G.C.Farlow Electric field enhancement of electron emission rates from Z1/2 centers in 4H-SiC, JAP, 106 (2009) 063702.

[132] L.Storasta and H.Tsuchida, Reduction of traps and improvement of carrier liofetime in 4H-SiC epilayers by ion implantation, Appl.Phys.Lett. 90, 062116 (2007). http://dx.doi.org/10.1063/1.2472530

[133] K.Danno, D.Nakamura, T.Kimoto, Investigation of carrier lifetime in 4H-SiC epilayers and lifetime control by electron irradiation, APL, 90 (2007) 202109. http://dx.doi.org/10.1063/1.2740580

[134] G.Alferi and T.Kimoto, Minority Carrier Transient Spectroscopy of As-grown, Electron Irradiated and Thermally Oxidized p-type 4H-SiC, Mater.Sci. Forum 778-780 (2014) 269-272. http://dx.doi.org/10.4028/www.scientific.net/MSF.778-780.269

[135] V. V. Makarov and N. N. Petrov, " Influence of the ion bombarding on cathodoluminescence of Silicon Carbide", Sov.Phys. Solid State 8, 1272-1276 (1966).

[136] V. V. Makarov, "Cathodoluminescence of different modification of the SiC crystals, containing radiation defects", Sov. Phys. Solid State 9, 457-461(1967).

[137] N. V. Kodrau and V. V. Makarov, " Luminescence spectrum of defects in ion implanted SiC", Sov. Phys. Semicond. 15, 813-815 (1981).

[138] L.Patrick and W.J.Choyke, Photoluminescence of Radiation Defects in Ion-Implanted 6H-SiC Phys.Rev. B 5 (1972) 3253.
http://dx.doi.org/10.1103/PhysRevB.5.3253

[139] V. M. Gusev, K. D. Demakov, V. M. Efimov, V. I. Ionov, M. G. Kosagonova, N. K. Prokofeva, V. G. Stolyarova, and Yu. N. Chekushkin, "electro- luminescence properties of several SiC polytypes, ion doped by Al", Sov. Phys. Semicond. 15, 1413-1416 (1981).

[140] V. V. Makarov and N. N. Petrov, "Cathodoluminescence of the silicon carbide mono crystals, irradiated by fast electrons", Sov.Phys.-Solid State 8, 2714-2715 (1966).

[141] V. V. Makarov, "Luminescence and optical properties of the neutron irradiated SiC", Sov. Phys. Solid State 13. 1974-1979 (1971).

[142] V.M. Gusev, K.D. Demakov, M.G., Gosagonova, M.B. Reifman, V.G. Stoliarova, Sov., Phys. Semicond. The study of electroluminescence JRC crystals, ion-doped with boron, aluminum and gallium, 9, 1238 (1975)

[143] K. D. Demakov, V. S.Ivanov, V. G. Stolyarova, and V. M. Tarasov, " Transition electroluminescence characteristics of the lights emitted diodes, formed on a-SiC by ion implantation of Al+ ions", Sov. Phys. Semicond. 12, 644-647 (1978).

[144] Yu. A. Vodakov, G. A. Lomakina, E. N. Mokhov, M. G. Ramm, and V. I. Sokolov, " Influence of the growth condition on thermal stability of the defect electroluminescence with D1 spectrum in 6H SiC irradiated by neutrons", Sov. Phys.Semicond. 20, 1347-1351 (1986).

[145] Yu. M. Suleimanov, A. M. Grekhov, and V. M. Grekhov, "Electron- oscillation structure of the D1 spectrum in irradiated silicon carbide", Sov. Phys. Solid State 25, 1060-1063 (1983).

[146] Yu. A. Vodakov, A. I. Girka. A. 0. Konstantinov, E. N. Mokhov, A. D.Roenkov, S. V. Svirida, V. V. Semenov, V. I. Sokolov, and A. V. Shishkin, "Ligth emitting diodes based on silicon carbide, irradiated by fast electrons", Semiconductors 26, 1041-1044(1992).

[147] Yu.A.Vodakov, A.A.Vol'fson, G.V.Zaritskii, E.N.Mokhov, A.G.Ostroumov, A.D.Roenkov, V.V.Semenov, V.I. Sokolov, and V.E. Udal'tsov Effective green light-emitting diodes based on silicon carbideSov.Phys. Semicond 26 (1992) 59.

[148] Lebedev A.A., B.Ya. Ber, N.V.Seredova, D.Yu Kazantsev, V.V.Kozlovski, Radiation-stimulated photoluminescence in electron irradiated 4H-SiC, Journal of Physics D: Applied Physics, 48, 485106 (2015). http://dx.doi.org/10.1088/0022-3727/48/48/485106

[149] A.A. Bergh, P.J. Dean Lightemitting diodes / (Oxford, 1976)

[150] R.G. Wilson, F.A. Stevie, and C.W. Magee, Secondary Ion Mass Spectrometry. A Practical Handbook for Depth Profiling and Bulk Impurity Analysis, Wiley, 1989.

[151] V.V.Kozlovski, A.A.Lebedev, E.V.Bogdanova, N.V.Seredova, Effect of irradiation with MeV protons and electrons on the conductivity compensation and photoluminescence of moderately doped p-4H-SiC (CVD) Semiconductors, 49, no. 9 (2015) 1163. http://dx.doi.org/10.1134/S106378261509016X

[152] J. Isoya, T. Umeda, N. Mizuochi, N.T. Son, E. Janzén and T. Ohshima, EPR Identification of Defects and Impurities in SiC: To be decisive, Materials Science Forum Vols. 600-603 (2009) pp 279-284. http://dx.doi.org/10.4028/www.scientific.net/MSF.600-603.279

[153] H.M.Ayedh, V.Bobal, R.Nipoti, A.Hallen, and B.G.Svensson, Formation of carbon vacancy in 4H silicon carbide during high-temperature processing, J Appl.Phys 115, (2014) 012005 http://dx.doi.org/10.1063/1.4837996

[154] V.Kozlovski, and V.Abrosimova, Radiation Defect Engineering, Selected Topics in Electronics and Systems,Vol.37 (World Scientific, Singapore, 2005).

[155] N.Iwamoto, B.G.Svensson. Point Defects in Silicon. Chapter Ten In: Defects in Semiconductors Ed. By L.Romano, V.Privitera and Ch.Jagadish. (Ser. Semiconductors and Semimetals. Vol. 91). Elsevier, 2015.

[156] A.A.Lebedev and V.V.Kozlovski, Irradiation of sublimation-grown p-SiC with 0.9-MeV electrons,Techn.Phys.Lett 40 (2014) 651. http://dx.doi.org/10.1134/S1063785014080094

[157] V.V.Kozlovski, A.A.Lebedev, V.N.Lomasov, E.V.Bogdanova and N.V.Seredova,Conductivity compensation in n-4H-SiC (CVD) under irradiation with 0.9 MeV electrons, Semiconductors, 48 (2014) 1006. http://dx.doi.org/10.1134/S1063782614080156

[158] W. F. Koehl, B. B. Buckley, F. J. Heremans, G. Calusine and D. D. Awschalom, Room temperature coherent control of defect spin qubits in silicon carbide, Nature, 479 (20110 84-88.

[159] A. G. Smart, Silicon carbide defects hold promise for device-friendly qubits, Physics Today,65 (2012) 10-11. http://dx.doi.org/10.1063/PT.3.1410

[160] P.G.Baranov, A.P. Bundakova, A.A.Soltamova, S.B.Orlinskii, I.V.Borovykh, R.Zondervan and J.Schmidt, Silicon vacancy in SiC as a promising quantum system for aingle-defect and single-photon spectroscopy, Phys.Rev B 83, 123203 (2011). http://dx.doi.org/10.1103/PhysRevB.83.125203

[161] T. C. Hain, F. Fuchs, V. A. Soltamov, P. G. Baranov, G. V. Astakhov, T. Hertel,and V. Dyakonov, Excitation and recombination dynamics of vacancy-related spin centersin silicon carbide, JOURNAL OF APPLIED PHYSICS 115, 133508 (2014). http://dx.doi.org/10.1063/1.4870456

[162] P.G.Baranov, V. A. Soltamov, A.A.Soltamova, G. V. Astakhov, and V. V. Dyakonov, Point defects in SiC as a promising basis for single-defect, single-photon spectroscopy with room temperature controllable quantum states, Mater.Sci.Forum 740-742, (2013) 425-430. http://dx.doi.org/10.4028/www.scientific.net/MSF.740-742.425

[163] V. Dyakonov, H.Kraus, V. A. Soltamov, F. Fuchs, D.Simin, S.Vaeth, A.Sperlich, P.Baranov, and G. V. Astakhov, Atomic-scale defects in silicon carbide for quantum sensing application, Mater.Sci.Forum 821-823, (2015) 355-358. http://dx.doi.org/10.4028/www.scientific.net/MSF.821-823.355

[164] V. A. Soltamov, D. O. Tolmachev, I. V. Il'in, G. V. Astakhov, V. V. Dyakonov, A. A. Soltamova, P. G. Baranov, Point defects in silicon carbide as a promising basis for spectroscopy of single defects with controllable quantum states at room temperature, Physics of the Solid State, 57, pp 891-899, (2015) http://dx.doi.org/10.1134/S1063783415050285

[164] V.S.Balandovich, Deep-level transient spectroscopy of radiation induced levels in 6H-SiC, Semiconductors, 33, p 1188 (1999). http://dx.doi.org/10.1134/1.1187846

CHAPTER 4

Effect of irradiation on the properties of SiC and parameters of SiC – based devices

Abstract

Results obtained in studying the effect of ionizing radiation on devices based on silicon carbide are considered. The analysis of published data illustrates the effect of intrinsic defects and radiation defects in the SiC crystal lattice on the properties of the epitaxial layers themselves, such as their doping level and polytype homogeneity. The radiation hardness of silicon carbide and silicon are compared. A conclusion is made that comparison of the values of charge carriers removal rate, obtained at room temperature for SiC and Si, is not fully adequate from the physical standpoint.

Keywords

Devices Radiation Hardness, Polytype Transformation, Reverse Recovery Time, Particle Detectors, Annealing

Contents

4.1 Change in parameters of SiC devices under irradiation

4.1.1 Schottky diodes

As already noted (see Chapter 1), silicon carbide is a highly promising materials for the development of devices on its basis for high-power and high-temperature electronics. At present, Schottky diodes (SDs) with breakdown voltage of ≥ 1200 V are manufactured on the basis of 4H-SiC. These devices can replace silicon-based pn diodes. Compared with silicon-based p-i-n diodes, SiC SDs show a substantially higher operation speed (because there is no injection of minority carriers) and lower reverse currents (due to the wider energy gap). The contact potential difference is approximately the same (~1 V) in both kinds of structures.

In addition, SiC SDs can be used as solar-blind UV photodetectors and detectors of charged particles. After the technology of III-N compounds and their solid solutions was developed, the UV photodetector is the only SiC optoelectronic device of practical interest. The reason is that light with hv > Eg is absorbed in GaN, which is a direct-gap semiconductor, near the surface. As a result, a considerable part of electron-hole pairs being formed are lost due to the surface recombination, with the photosensitivity of the detector becoming lower. In silicon carbide, which is an indirect-band semiconductor, light penetrates much deeper and the effect of the surface recombination on the lifetime of carriers being formed is substantially weaker.

In [1], SDs fabricated by deposition of Pt-W-Cr onto 6H-SiC samples grown by the Lely method with Nd-Na = 10^{16} -5 10^{17} cm^{-3} were studied. The structures retained their electrical parameters up to temperatures of ~450 °C. It was shown that the thermal stability is due to the use of a multilayer metallic composite that provides the stability of metal/silicon carbide interfaces. It was also found that the characteristics of the samples under study were irreversibly changed upon a combined effect of fast neutrons and accompanying radiation at doses of 4.42 neutron/cm^2 and 8.67 10^5 R, respectively. The degradation of the characteristics was the stronger, the lower the initial doping level of the material.

In [2], characteristics of 4H-SiC SDs irradiated with 7.0 MeV C ions at fluence in the range 10^9-10^{10} cm^{-2} were studied. It was found that the irradiation did not lead to a rise in reverse currents, but caused an increase in the series resistance, which was apparently due to the larger degree of SiC compensation (Fig. 4.1). The temperature dependence of carrier mobility determined from I-V measurements in the temperature range 100-700 K shows a T^{-3} behavior of the mobility, determined for both unirradiated and irradiated diodes. The authors of [2] also noted that the ion irradiation results in that the epilayer resistance grows due to the deactivation of nitrogen at the end of the ion range, where the concentration of defects created by the ion beam is higher. A deep insulating layer formed at the end of the range remained stable upon annealing at 700 K for 1 h.

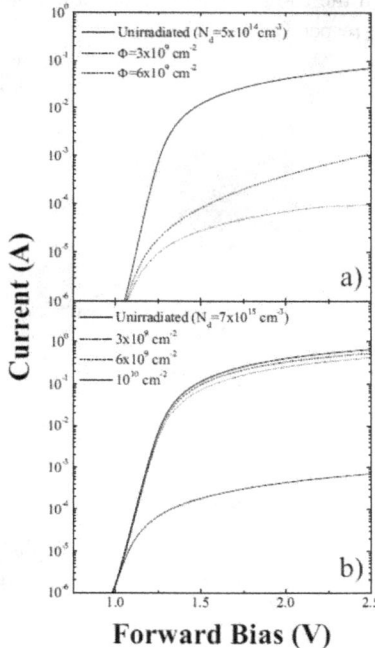

Fig. 4.1. Forward I-V characteristics of 4H-SiC Schottky diodes before and after the irradiation with 7.0 MeV C ions at different fluences for epilayers doped to (a) Nd = 5.0×10^{14} cm^{-3} and (b) Nd = 7.0×10^{15} cm^{-3} [2].

The effect of irradiation with electrons from Sr-90 radio nuclide (fluence 6×10^{14} electron/cm^2) on similar 4H-SiC SDs was studied in [3]. It was noted that the current-voltage characteristics of the SDs are well described in terms of the thermionic emission (TE) in the temperature range of 120-300 K, but depart from the TE theory at temperatures below 120 K. The current flowing across the interface at a bias of 2.0 V from pure thermionic emission to thermionic field emission within the depletion region with the free carrier concentrations of the devices decreased from $7.8 \cdot 10^{15}$ to $6.8 \cdot 10^{15} cm^{-3}$ after the HEE irradiation. The modified Richardson constants were found from the Gaussian distribution of the barrier height across the contact to be 133 and 163 A $cm^{-2}K^{-2}$ for as-fabricated and irradiated diodes, respectively.

In [4], 4H-SiC Schottky diodes manufactured by CREE Inc. (United States) with breakdown voltages of 600 and 1200 V were studied. The uncompensated donor impurity concentration (Nd - Na) in the as-fabricated devices before irradiation was $\sim 6.5 \times 10^{15}$ cm^{-3} and $\sim 4.5 \times 10^{15}$ cm^{-3}, respectively.

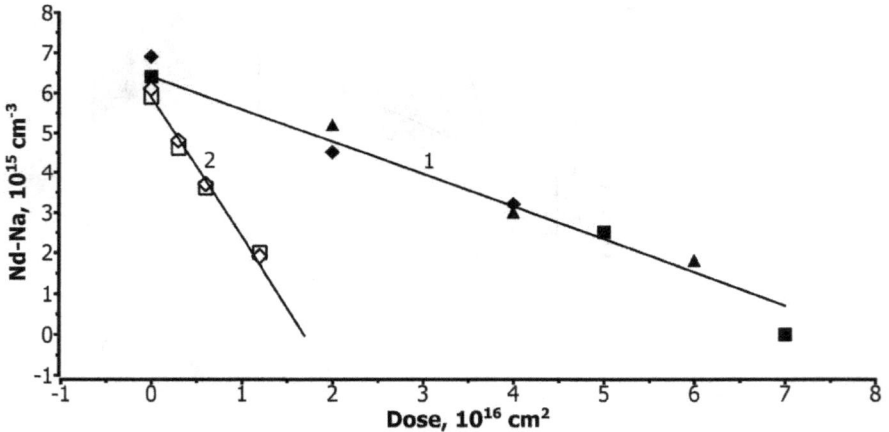

Fig. 4.2. Dependence of Nd-Na on the irradiation dose for the case of the electron irradiation with energies of (1) 0.9 MeV and (2) 3.5 MeV. Different symbols correspond to different diodes. Adapted from [4].

The irradiation with 0.9 MeV electrons was performed on a resonant transformer accelerator. The uncompensated acceptor concentrations (Na - Nd) in the as-fabricated and irradiated samples were found from capacitance-voltage (C-U) characteristics.

A linear decrease in Nd - Na was observed in the course of experiments for both types of Schottky diodes (Fig. 4.2). The carrier removal rate upon irradiation with 0.9 MeV electrons was Vd ≈ 0.096 cm^{-1} at the concentration of 6.5×10^{15} cm^{-3} in the base region of the Schottky diode and Vd ≈ 0.073 cm^{-1} for the concentration of 4.5×10^{15} cm^{-3}. This value is approximately twice [5] smaller than that for silicon with the same Nd - Na. If a comparison is made with a silicon p-i-n diode having the same breakdown voltage, it is necessary to take Nd - Na two orders of magnitude lower than that in SiC because the critical breakdown field in Si is 10 times lower than that in SiC. This means that an approximately 200 times higher irradiation dose is required for compensation of a SiC Schottky diode, compared with the compensation of a Si p-i-n diode with the same breakdown voltage.

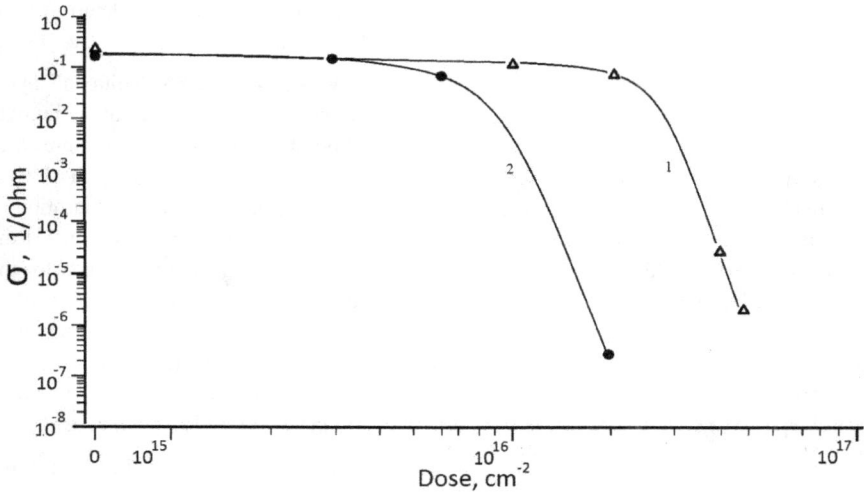

Fig. 4.3. Forward conductivity of Schottky diodes on the irradiation dose for the case of irradiation with (1) 0.9 MeV and (2) 3.5 MeV electrons. Adapted from [4].

Figure 4.3 shows how the conductivity of the SDs under study varies with the irradiation dose. It can be seen in the figures that the conductivity of the samples remains unchanged at irradiation doses of $\sim 1 \times 10^{17}$ cm^{-2}, although C-U measurements demonstrate that at these doses Nd - Na → 0. In the given scheme of the experiment, the experimentally measured conductivity (σ) can be written as

$$\sigma = (Rb + Rsub + Rc)^{-1} \tag{4.1}$$

where Rb is the resistance of the base region (thin lightly doped layer); Rsub is the substrate resistance, and Rc is the contact resistance.

Apparently, the last term (Rc) does not change significantly under irradiation. The doping level of the SiC substrate is approximately three orders of magnitude higher than that of the epitaxial layer. For the irradiation type used and the dose of $<2\times10^{17}$ cm^{-2}, not more than 1% of carriers present in the substrate will be removed. Thus, the value of Rsub must not strongly increase, either. Under irradiation, Rb will grow due to the decrease in the concentration of carriers and in their mobility. However, even when the experimentally measured Nd - Na approaches zero, the carrier concentration in the base will not be zero. These carriers will be excited from deep donor centers (including those appearing upon irradiation). It is the presence of these carriers that will determine the conductivity of the system at irradiation doses of $\sim1\times10^{17}$ cm^{-2}.

A comparison of the electrical characteristics of CVD-grown 4H-SiC epitaxial layers upon irradiation with 0.9 and 3.5 MeV electrons demonstrated that the donor removal rate increases by nearly a factor of 4 as the energy of bombarding electrons becomes four times higher, despite that the formation cross-section of primary radiation defects (Frenkel pairs in the carbon sublattice) responsible for the conductivity compensation of the material is nearly energy-independent within this range. It was suggested in [6] the influence exerted by the energy of primary knock-on atoms (PKAs) is the reason for the observed differences. First, cascade processes become important with increasing PKA energy. Second, the average distance between genetically related Frenkel pairs increases, and, as a consequence, the fraction of defects that do not recombine under irradiation becomes larger. In [6], the recombination radius of a Frenkel pair in the carbon sublattice and the possible charge state of the recombining components were also evaluated.

Comparison of Figs. 4.2 and 4.3 also shows that, at doses of $\sim5\times10^{16}$ cm^{-2}, the conductivity of the samples decreases by five orders of magnitude, whereas the experimentally measured Nd - Na becomes only three to four times lower. A possible reason is that n < Nd - Na because of the high ionization energy of the main donor levels in SiC. With increasing degree of compensation, the difference between n and Nd - Na will grow. Another reason for the so significant decrease in conductivity may be associated with the decrease in the carrier mobility. Presumably, this effect invites further studies.

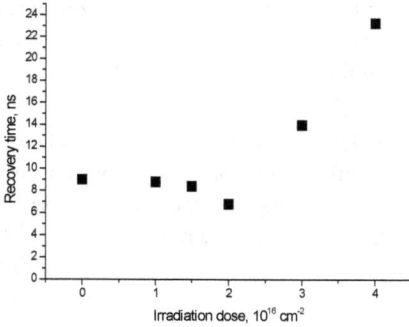

Fig. 4.4. Dependence of the reverse recovery time of the diode on the dose of irradiation with 0.9 MeV electrodes.

The reverse recovery time (τ_{REC}) measurements were made on similar SDs [7] on a Lemsys Inc. Diode Measurement System (DMS) with a QPR 100A Turn-off unit. This was done by the method of active switch-off of a diode from 20 A on applying a reverse voltage of 100 V, with a current decay front (dI_F/dt) of 100 A/μs, upon irradiation with various doses (Fig. 4.4). It can be seen in the figure that the operation speed of the diodes is slightly improved up to doses of $\sim 3 \times 10^{16}$ cm^{-2}, and, as the dose increases further, this parameter is sweepingly deteriorated. The observed dependence can be explained as follows [7].

It is known that the recharge time of barrier capacitance (τ_M) between the two equilibrium states mostly determined by the Maxwell relaxation time expression (4.2):

$$\tau_M = \rho \varepsilon \varepsilon_0 \qquad\qquad (4.2)$$

where ρ is the resistivity, ε the dielectric constant, and ε_0 the electric constant. Rewriting the expression as $\tau_M = R_s C_0 \varepsilon \varepsilon_0$, where R_s is series on resistance and C_0 - barrier capacitance at zero bias, we can see that reverse recovery time (τ_{rr}) determined by τ_M is equal to $R_s C_0$.

Thus, the change in the reverse recovery time will be determined by the influence of two mutually compensating processes. Specifically, the barrier capacitance becomes lower due to the decrease in the N_d - N_a concentration, on the one hand, and to the increase in R_s because of the decrease in the number of carriers and in their mobility, on the other hand.

123

The latter occurs because the carrier lifetime becomes shorter due to the scattering on additional radiation defects. As a result, the decrease in τ_{rr} at low irradiation doses, when the increase in R_s is not significant yet, is determined by the decrease in C_0. At doses > 2×10^{16} cm^{-2}, the key influence on the variation of τ_{rr} will be exerted by the increase in R_s and ohmic resistance of n-base.

In [8-10], the degradation of SiC UV photodetectors under irradiation with neutrons and ions was studied. In [8], the effects of ion irradiation on the response of UV photodetectors were studied by monitoring the spectral response as a function of the irradiation dose. The devices irradiated with a beam of 1, 4, and 10 MeV Si$^+$ ions show a change in the response, depending on the energy of irradiating ions. The unexpected huge optical effect, compared to negligible influence on the reverse bias leakage current, was correlated with the nature of irradiation-induced damage and with its position within the optically active layer of the device.

In [9], the effect of irradiation with 1 MeV neutrons was considered. It was shown that the photoresponse is slightly affected by the irradiation up to the threshold fluence $\Phi_{critical}$ = 8 x 10^{14} cm^{-2}. At fluences $\geq$ $\Phi_{critical}$, the rejection rate is in the order of 10^3 in the range 200-320 nm, and less than 10^2 at about 320 nm.

Studies of the effect of heavy ions (Xe) on characteristics of 4H-SiC UV photodetectors enabled the authors of [10] to formulate the following conclusions:

- The short-wavelength part of the photosensitivity spectrum ($\lambda \leq 250$ nm) of 4H-SiC UV photodetector structures with Schottky barriers is determined by the surface perfection of the semiconductor. It is quasi-strained due to the surface imperfections (and, in particular, fluctuation traps). Irradiation of the detectors with high-energy Xe ions at a fluence of 6×10^9 cm$_{-2}$ reveals this dependence to a greater extent.

- UV detector structures remain operable under irradiation with 167 MeV heavy Xe ions at fluence of 6×10^9 cm^{-2} (Fig. 4.5).

As the working temperature of the detectors is raised to 180 °C, their photosensitivity is preserved and even an increase is observed in the endurance and burn-up life of 4H-SiC devices at elevated temperatures.

The interest in SiC as a starting material for nuclear detectors is due to its supposed high radiation hardness and possibility of operation of devices at elevated temperatures. The detectors based on a pn structure or an SD are, in fact, an ionization chamber fabricated in the solid state.

Until quite recently, a severe obstacle to designing detectors has been the poor quality of epitaxial films. The recent progress is associated with the appearance of films with thickness comparable with the deceleration length of short-range ions. For example, a space-charge layer with a thickness of tens of micrometers can be developed even at an Nd - Na concentration of $\sim 10^{15}$ cm^{-3} and hole diffusion length of 2-3 μm because of the high critical electric field strength in SiC. The low concentration of nonradiative recombination centers, achieved in layers of this kind, provides carrier lifetimes sufficiently high for the effective transport.

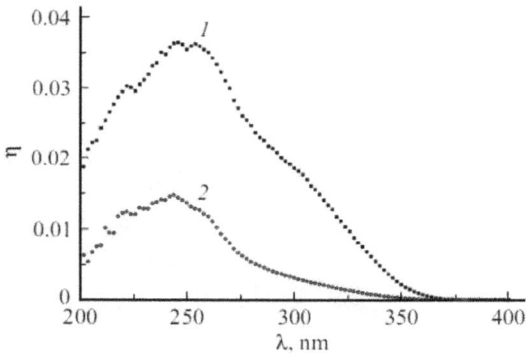

Fig. 4.5. Spectral dependences of the quantum efficiency of photoconversion by 4H-SiC photodetectors with Schottky barriers: (1) as-fabricated sample and (2) sample irradiated with 167 MeV Xe ions at fluence of 6×10^9 cm^{-2}. Measurement temperature was 25 °C [10].

The conversion efficiency of the particle energy into the nonequilibrium charge is determined by the average energy (ε) consumed for generating an electron-hole pair. Pairs are generated in collision cascades produced by secondary fast electrons. In addition, the energy of hot carriers is expended for excitation of phonons. There are published empirical formulas that relate the energy gap width Eg, phonon loss Eph, and resulting value of ε of the type ε = 2.16Eg + Eph [11, 12]. The values of Eph are close values for various materials, being 1.2 eV for diamond and GaAs. With this value, the factor multiplying Eg was found to be ~1.9 in [13]. In other words, the ionization occurred in SiC films under study in [13] with a somewhat higher efficiency, compared with the materials mentioned above.

The typical dependence of the average signal amplitude on the voltage across the detector is shown in Fig. 4.6. The dependence has two portions that differ in the rate at which the amplitude increases, which correspond to the voltages before and after the complete depletion of the structure. Indeed, while the electric field region occupies only a part of the n-base, the diffusion of holes in the base starts to contribute to the carrier transport in addition to the carrier drift in the nonzero field region.

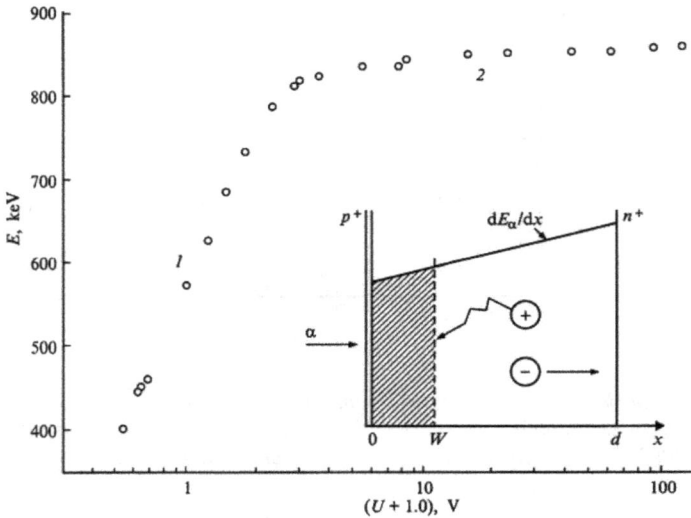

Fig. 4.6. Dependence of the average signal amplitude on the reverse voltage applied to the detector. The portions before the depletion (1) and after the complete depletion (2) of the structure are indicated. The inset shows the configuration used in the diffusion analysis; the nonzero field region W is shaded. [13]

The slower, compared with the drift, diffusion rate results in a considerable loss of charge via the carrier recombination. To a first approximation, the signal grows in proportion to the width of the nonzero field region.

When the depletion is reached and the nonzero field region extends to the whole lightly doped film, there only occurs the carrier drift. Because electrons and holes are effectively separated by the field, the loss of charge is determined in this case by the carrier localization on trapping centers, rather than by the recombination. In this case, centers are manifested which retain carriers during a time exceeding that in which a pulse is formed by the recording equipment (on the order of microseconds). The signal amplitude grows with increasing reverse voltage due to the rise in the drift velocity. The charge transfer can be considered complete in the saturation portion.

In [14-21], the results obtained in detection of various kinds of radiation with SiC SD detectors were considered. In [14, 15], detector arrays constituted by 22 SiC SDs with diameters of 200 and 400 $\mu\mu$ were used to record neutrons, gamma photons, and α-particles. It was noted that the detectors expectedly exhibited weak sensitivity to gamma radiation, with the signal linearly increasing with the dose rate. No effect of temperature on the detector characteristics was observed in the range 22-89 °C. A conclusion was made that SiC is a promising material for development of detectors intended for use at elevated temperatures and/or high radiation fields.

In [16], n-4H-SiC SDs were used to detect 22 MeV electrons and 6 MeV protons. It was found that they exhibit a linear variation of the electron-induced collected charge with electron dose within the range 1-10 Gy and a linear trend in the range 2-7 Gy/min. The sensitivity per volume is almost a factor of 2 lower than that of the standard silicon dosimeters, but it is well above that of the best available commercial CVD diamond.

The potential of SiC for detection of X-rays from 241AM was considered in [17, 18]. SDs with area of ~3 mm^2 were fabricated from 4H-SiC with the use of gold and titanium. The best devices had leakage currents in the range 2-18 pA/cm^2 at room temperature. The authors of [17] noted that the low noise level of SiC detectors at room temperature enables their effective use without cooling. This makes it possible to reduce the end value of a device and can compensate for the high cost of epitaxial SiC structures.

In [19,] samples were characterized, chosen among several batches of epitaxial 4H-SiC Schottky diodes with different effective nitrogen doping levels and layer thicknesses. The nominal depletion width of the devices ranges from 20 to 40 micrometers, and the effective doping concentrations, from below 10^{14} cm^{-3} to about 10^{15} cm^{-3}. The epilayer thicknesses and the concentration gradients were determined by capacitance-voltage measurements. The charge collection efficiency (CCE) was characterize with a minimum-ionizing β-source. The results of the C-V measurements are consistent with the CCE characterization. In particular, the two samples under study exhibited the 100% CCE, with the signal well separated from the pedestal and a reproducible response observed to minimum-ionizing particles.

The resolution of SiC detectors in spectrometry of α-particles with energy of 5.1-5.5 MeV has been studied [20, 21]. The detectors were fabricated from 26-μm-thick CVD-grown 4H-SiC films with Nd - Na $\approx 1 \times 10^{15}$ cm^{-3}. A resolution of 0.5% was achieved for SiC detectors for the first time, which made it possible to visually reveal fine-structure peaks of the α-spectrum (Fig. 4.7). A value of 7.71 eV was obtained for the average formation energy of an electron-hole pair.

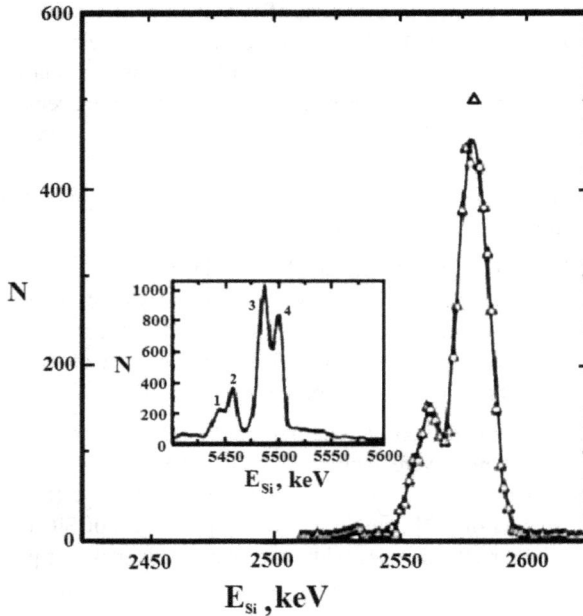

Fig. 4.7. α-Decay spectra measured with a SiC detector for the isotope mixtures ^{241}Am + ^{238}Pu. The inset shows the same spectra measured with a silicon detector. Energies (keV): 5443, 5456, 5485, and 5499 (for, respectively, lines 1- 4 in the lower panel) Adapted from [20].

In [22], the influence exerted on the characteristics of a 4H-SiC SD detector by irradiation with protons (8 MeV) and electrons (900 keV) at doses of 10^{14} and 3×10^{16} cm^{-2}, respectively, was studied. It was found that the proton irradiation leads to a monotonic fall of the transfer efficiency. In the case of the electron irradiation, the efficiency of the detectors remained unchanged at doses in the range $(1-3) \times 10^{16}$ cm^{-2}. However, the dose of 3×10^{16} cm^{-2} led to a substantially higher nonuniformity of the charge-transfer conditions over the sample volume.

The effect of annealing on the characteristics of SiC SD detectors irradiated with 1 MeV electrons has also been studied [23]. The annealing at 300 °C considerably improved the CCE parameter deteriorated after the irradiation. However, annealing at temperatures of 350 °C and more led to degradation of the devices. For example, the authors of [23] noted that the leakage current nearly unchanged upon irradiation increased by three orders of magnitude upon heating to 400 °C. It is noteworthy that these temperatures are limiting for the working capacity of SiC SDs because the barrier is thermally degraded and becomes ohmic.

4.1.2 PN diodes

In bipolar SiC diodes, to which PN diodes also belong, there is a degradation mechanism associated with the formation of 3C-SiC polytype inclusions within a structure based on the hexagonal polytype under the action of a forward current (see Section 4.2 for more detail). That is why SiC PN diodes are not commercially manufactured on a large scale, and fabricated devices are more frequently used in the reverse-bias mode, e.g., as detectors.

Another possible way to solve the problem of degradation of SiC PN diodes is by developing JBS (Junction Barrier Schottky) diode structures. The working area of these devices contains both PN junctions and SDs. The JBS characteristics are a combination of the characteristics of SDs and a PN diode. Under forward bias, the JBS operates as an SD, and the breakdown reverse voltage is close to the breakdown voltage of a PN diode. Because a forward-biased SD has no injection or electron-hole recombination in the base, JBS structures show no degradation associated with the transformation of the SiC polytype.

In [24], the effect of proton irradiation (energy 62.5 MeV, dose 5×10^{13} cm^{-2}) on the parameters of a 4H-SiC JBS diode (Nd - Na = 1×10^{15} cm^{-3}) was studied. Both deterioration of some characteristics (increase in the series resistance) and improvement of others (decrease in the leakage current and increase in the breakdown voltage) were observed. Presumably, all these effects occur because the degree of compensation of the base region of the device grows under irradiation.

The effect of irradiation with 5 MeV protons on SiC PN diodes was analyzed in [25]. This study was for the most part concerned with parameters of the RDs being formed. Therefore, low irradiation doses (10^{11} cm^{-2}) were used at a base concentration of 1×10^{15} cm^{-3}. The presence of the PN junction enabled a study by capacitive methods of the traps formed in both lower and upper halves of the energy gap of SiC. Also, a decrease in the hole lifetime as a result of irradiation from 0.78 to 0.16 μs was observed. The authors of [25] reasonably attributed this effect to the increase in the concentration of the $Z_{1/2}$ center.

The development and analysis of charge-particle detectors based on SiC PN diodes have been the subject of rather numerous publications (see, e.g., [26-29]). The operation physics of the detector is the same in the case of a PN junction or a detector based on an SD. In practical regard, the PN detector has higher working temperatures; also, lower leakage currents and higher breakdown voltage can be reached at the same base doping level.

In [30], specific features of the operation of diodes irradiated with 8 MeV protons at a dose 3×10^{14} cm^{-2} were studied. The detectors were fabricated on the basis of CVD-grown epitaxial 4H-SiC films (Nd - Na ≈ 10^{14} cm^{-3}, thickness 55 μm) manufactured by CREE Inc. [31]. A temporal instability of the principal characteristics of the detector, signal amplitude and resolving power, was observed. A conclusion was made that this effect is due to the long-term capture of nonequilibrium carriers by RDs and to the resulting appearance of a polarization voltage in the bulk of the detector.

In [32], detectors based on p-type 6H-SiC epitaxial layers were studied. Characteristics of a detector before and after the irradiation with 1 MeV electrons were measured. A CCE of 93% was obtained for non-electron-irradiated diodes, and no significant change in the CCE was observed for the diodes irradiated with electrons at fluence below 1×10^{15} cm^{-2}. A degradation of the CCE was observed after irradiation at fluence exceeding 5×10^{15} cm^{-2}.

A general analysis of the radiation hardness of charge-particle detectors based on wide-bandgap semiconductor has been made [33]. It was noted that a mutual relationship between three processes is observed in detectors of this kind upon introduction of radiation defects. The deep compensation predetermines the localization as the dominant kind of the capture, which, in turn, gives rise to the polarization voltage.

A situation was considered in which the carrier generation and capture processes are governed by radiation centers. For the simplest case of a single level, the optimal working temperature and the depth of the level excluded from the capture process were determined. A comparison of various materials demonstrated that only the temperatures are strictly individual, whereas the depth level related to the energy gap width has a common value close to 1/3 (at the center parameters specified in [33]).

The radiation hardness of SiC and GaN have been studied at CERN when effective detectors for Large Hadron Collider (LHC) were developed [34, 35]. The following conclusion was made on the basis of the results obtained: "The studies have shown that the relatively small signal generated by MIPs degrades faster in these materials than in silicon after hadron irradiation. As a consequence, they are not suitable for hadron collider applications due to this degradation of their charge collection properties, besides other practical considerations like cost and availability of adequate-size wafers (4 inches and 6 inches)" [34]. According to the data obtained, SiC detectors are inferior not only to diamond, but also to silicon. In our opinion, the conclusion that SiC is unsuitable for development of detectors seems to be somewhat premature. During the 5-7 years that have passed since the time when the studies considered in [34, 35] were carried out, the advances in the silicon carbide technology have continued. For example, SiC structures with diameters of 4 to 6 inches are already available on the market.

The conclusion made in [34, 35] can be divided into two parts. The first is the low sensitivity of a detector, and second is its radiation hardness proper. As for the second factor, is that SiC degrades more slowly than diamond and most types of silicon detectors. The originally low sensitivity of SiC could be due to the poor crystal perfection of the material 5-7 years ago. What is more, the properties of SiC were not optimized, in contrast to Si, for being used as a detector. At the same time, studies aimed to create radiation-hard silicon have been carried out during more than one decade. The radiation hardnesses of SiC and Si are compared in more detail in Section 4.3. On the whole, it can be concluded that the question of whether or not SiC is promising for development of nuclear detectors requires further analysis.

4.1.3 SiC field – effect transistors

Another class of unipolar devices that are being actively developed on the basis of SiC are field-effect transistors (FETs). Depending on the gate fabrication technology, these devices are subdivided into JFETs (gate in the form of a pn junction), MESFETs (gate in the form of an SD, and MOSFET (MIS structure as the gate).

In [36], a 4H-SiC JFET was irradiated with various kinds of charged particles: 6.8 Mrad gamma photons from a ^{60}Co source; 4 MeV protons with a dose of 9×10^{11} cm^{-2}, and 63 MeV protons with a dose of 5×10^{13} cm^{-2}. Prior to irradiation, the transistor had a blocking voltage of 600 V and a rated forward current of 2 A. After the irradiation with gamma photons and 4 MeV protons, a slight increase in the device resistance was observed in the ON-state and an insignificant decrease in leakage currents in the OFF-state. The irradiation with 63 MeV protons led to a pronounced decrease in the forward current.

The effect of 1 MeV neutrons and gamma photons on 4H-SiC MESFETs and SDs at room temperature was examined in [37]. The highest neutron flux and total gamma-ray dose were 1×10^{15} cm^{-2} and 3.3 Mrad, respectively. After a neutron flux of 1×10^{13} cm^{-2}, the current characteristics of the MESFET were only slightly changed, and the Schottky contacts of the gate contacts and of the Ni, Ti/4H-SiC Shottky Diodes showed no obvious degradation. With the neutron flux raised further, the drain current of the SiC MESFET decreased and its threshold voltage increased. The value of ϕ_B of the Schottky gate contact decreased when the neutron flux was more than, or equal to 2.5×10^{14} cm^{-2}. On the whole, a conclusion was made that the damage to the SiC Schottky interface and the radiation defects in the bulk material are the main mechanisms of the performance degradation of the experimental devices, and a high doping level of the active region will improve the tolerance to the neutron irradiation.

The MOSFET is believed to be one of the most promising SiC-based FETs as the highest power device. The role of the gate insulator is commonly played by SiO_2, which is formed, similarly to the case of silicon, on the semiconductor surface by oxidation. However, in contrast to Si, the number of surface states formed at the SiC/SiO$_2$ interface is an order of magnitude larger. Another, still unsolved technological problem is the low carrier mobility in the inversion layer. Extensive studies are still being carried out, aimed to further develop the technology of SiC formation for MOSFETs and optimize its parameter.

The effect of irradiation on the parameters of this device was also studied in [37-42]. As the material served various SiC polytypes: 3C [38], 6H [39, 40], and 4H [41]. Gamma irradiation was used [38-40], and RDs that appeared after the transistor channel was formed by implantation of N were examined. The formation of traps both at the SiC/SiO$_2$ interface and deep within the oxide layer was recorded. Presumably, SiO_2 has lower radiation hardness, compared with silicon carbide, and its degradation begins earlier. In our opinion, these studies are not directly associated with an analysis of radiation defects in SiC.

To conclude this section, it can be stated that the degradation of SiC under irradiation is associated, similarly to the case of other semiconductors, with the formation of RDs in the bulk of the material. As a result, the forward resistance of the device increases and, at the maximum irradiation dose, the semiconductor becomes a poor insulator.

Thus, the degradation rate of the given device will be determined by the relationship between the carrier removal rate at a given kind of irradiation and the value of Nd - Na (Na - Nd) in the most lightly doped region of the device. Therefore, it would be expected that high-voltage diodes and field - effect transistors degrade most rapidly, and low-voltage Zener diodes and tunnel diodes do so more slowly.

An interesting specific feature of the effect of radiation on SiC diodes is the improvement of device parameters (decrease in leakage currents) at low irradiation doses, more than once observed experimentally. Probably, this is due to the band structure of silicon carbide as a wide-bandgap semiconductor. The concentration of intrinsic carriers, which determines reverse currents in the "ideal" pn structure, is 10^{-6} cm^{-3} for 6H-SiC. That is the currents due to these carriers are below the measurement sensitivity limit of modern apparatus. Therefore, all the reverse currents recorded at room temperature are associated with shallow defects or impurity centers. These levels are presumably compensated in the initial stages of irradiation, and the leakage currents become lower.

On the whole, the increase in leakage currents is rather rarely mentioned in the case of SiC irradiation. This is only observed at large doses. Because of the wide energy gap of SiC, the main RDs have a substantial ionization energy and the time of their recharging is long. For example, the recharging time of an RD characteristic of 6H-SiC, R center, is ~1-6 s. Therefore, the current due to the recharging of RDs at room temperature will be only significant at their high concentration (large irradiation dose).

Another important specific feature of the radiation hardness of SiC devices, mentioned in a number of publications, consists in that it grows with increasing temperature at which the irradiation is performed. This is due to the elimination of primary radiation defects by annealing. This problem is considered in more detail in Section 4.3.

4.2   Possible transformation of the SiC polytype under irradiation

4.2.1   Possible resons for the polytypism of SiC

The existence of a large number of polytypes of silicon carbide makes this material rather promising for development of various types of heterostructures and device structures on their basis.

In [43], the prospects for possible application of 6H-SiC/3C-SiC heterostructures in High Electron Mobility Transistors (HEMT) were analyzed. It was found that (as also in AlGaN/GaN structures) polarization effects play an important role in the electron accumulation near the interface. It was found that the sheet electron density in the active layer of the 6H-SiC/3C-SiC heterostructure markedly increases after a spontaneous polarization appears. For a gate length Lg =200 nm, the intrinsic cutoff frequency f_t was estimated to be 160 GHz, with a slight decrease in the presence of the polarization. (For an $Al_{0.15}$ $Ga_{0.85}$ N/GaN HEMT having the same configuration and doping level, the authors [43] calculated nearly the same f_t, but with a more pronounced decrease at negative gate voltages for the case when the polarization was taken into account.) Finally, it was concluded that 6H-SiC/3C-SiC heterostructures can be regarded as a promising candidate for development of HEMTs.

In [44], results of a simulation study of the behavior of a 3C-SiC bipolar transistor with a 6H-SiC heterojunction emitter were reported. It was shown that, despite the potential barrier originating from the conduction band offset, *npn* devices offer a higher common-emitter gain in comparison with *pnp* devices. The base voltage corresponding to the maximum gain of the device is controlled by the carrier tunneling across the collector-base junction and is different for *npn* and *pnp* devices. Thus, the development of a technology for creating structurally perfect SiC heterostructures seems to be an interesting research task.

As noted in the preceding sections of the book, native defects of the crystal lattice of SiC exert an exceedingly strong influence on its properties and parameters of SiC devices. In this section, we consider the effect of native defects on the polytypism of silicon carbide.

At present, there is no theory that would be satisfactory in every respect in explaining why SiC and some other materials crystallize in a wide variety of polytypes. It is not completely clear, either, what factors favor the formation of a particular polytype.

It was observed in [46 - 48] that introduction of certain impurities into the growth zone of SiC layers can yield epitaxial films having a polytype other than that of the substrate used. For example, the introduction of rare-earth elements Sc and Tb, and also Al and B, led to growth of 4H-SiC films on 6H-SiC substrates. The most effective transformation of a growing layer, 6H $\Rightarrow$ 4H, was observed upon introduction of Group-IV impurities Sn, Pb, and Ge. Group-V impurities (nitrogen and phosphorus) favored growth of the 3C polytype.

It was also found that changing the Si/C ratio in the growth zone strongly affects the heteropolytype epitaxy. For example, an increase in the Si concentration made more probable formation of 3C SiC or other polytypes with a low percentage of hexagonality. At the same time, the introduction of an excess carbon enabled growth of 4H-SiC epitaxial layers on 6H substrates also from Gd and Dy melts [48]. It was also noted in [46] that the transformation of the substrate polytype occurs the most easily in growth in the (0001)C direction. In this case, the growth temperature and growth rate affected the process of heteropolytype epitaxy only slightly.

It was reported in [49, 50] that thick epitaxial 4H-SiC layers were obtained on 15R- and 6H-SiC substrates, with these structures then used as seeds for growth of single-crystal ingots. The growth was performed on the (0001)C face, with introduction of Sc into the vapor phase. This technique can produce epitaxial layers with both p- and N-types of conduction. It was noted [50] that a high Sc concentration exceeding 10^{17} cm^{-3} in epitaxial films gave rise to mechanical stresses in the films. On the whole, the thus obtained epitaxial layers had a rather high structural perfection, which enabled fabrication of JFET transistors on their basis [51].

As the nature of polytypism has not been elucidated yet, it is rather difficult to understand the nature of heteropolytype epitaxy, too. Presumably, not only the impurity composition of the growth zone, but also a number of other factors, from thermodynamic (pressure, temperature) to crystallographic (orientation and defectiveness of a substrate), can affect the probability of polytypic epitaxy. For example, it was observed in [52] that use of a 6H-SiC Lely substrate with a high dislocation density ($\sim 10^5$ cm^{-2}) in the standard (employed to grow 6H layers) epitaxial process leads to growth of 3C epitaxial layers.

Several the most efficient theoretical approaches can be distinguished among those presently existing. Those based on Frank's dislocation theory [53-55] were aimed to confirm the possibility of existence of a large number of polytypic forms and their relatively high stability, but disregarded the effect of real experimental conditions (temperature, pressure, and composition of the growth zone) on the polytypism. In a number of other studies, the main attention was given to a thermodynamic analysis of the growth of SiC polytypes (see [4] and refs. therein). Recent results were discussed in detail in [56]. According to [56], this approach can account for the predominant growth of the 4H-SiC polytype at an excess of C, growth of 3C-SiC films at an excess of Si, and a number of other experimental results. At the same time, this approach disregards the effect of impurities and structural defects in the substrate on the heteropolytype epitaxy.

In [46], processes of heteropolytype epitaxy were related to the stoichiometric composition of various SiC polytypes. It was assumed that introduction of carbon vacancies V_C, caused by the excess of silicon, leads to compression of the crystal lattice. In this case, cubic, rather than hexagonal orientation of layers becomes more energetically favorable. Thus, SiC polytypes can be regarded as phases with different SiC ratios and it can be assumed that an increase in N_V^C in a growing layer leads to a change of the polytype, making lower its hexagonality, i.e. $1 - \gamma \sim N_V^C$.

This approach was further developed in [57- 59]. In [58], experimental data on the effect of intrinsic defects in the crystal lattice on the polytypism of silicon carbide were analyzed. A simple analytical expression relating the hexagonality of a polytype to the concentration of carbon and silicon vacancies was derived. In [59], a model of transformation of SiC polytypes in the course of growth of an epitaxial layer, based on a time dependence of the concentration of carbon vacancies in the intermediate layer, was suggested and used to analyze the available experimental data. It was shown that the parameter $\eta = G\tau/L_T$ (where L_T is the thickness of the intermediate layer; G, film growth rate; and τ, lifetime of a vacancy in the intermediate layer) is independent of the method used to grow an epitaxial layer and of the growth temperature, being determined only by the concentration of carbon vacancies in the substrate and film.

The scheme of the process of heteropolytype epitaxy of SiC, presented in [58, 59], is strongly simplified and disregards numerous factors governing the film growth. At the same time, the approach suggested can, in our opinion, serve as a basis for analyzing the already available published data and for determining ways of further experimental research.

4.2.2 Selected experimental results

Heteropolytype structures have been fabricated by CVD method [60, 61], molecular-beam epitaxy [62], and sublimation [63, 64]. However, despite separate observations of quantum-confinement effects near the heterointerface, no technology for stable fabrication of SiC devices with heterojunctions has been developed.

Detailed studies of the transition region between two polytypes have shown [65-68] that it is constituted by alternating bands with structure of both polytypes. In [69], an attempt was made to supplement the previously suggested model [47, 48] by using the spinodal decomposition concept.

In the case of a fast phase transition, one (or both) of the coexisting phases is unstable. In this case, a spinodal decomposition of the system may occur, accompanied by the enhancement of accidental particle concentration nonuniformities [70]. Then, appearance of density-modulated relaxation structures becomes possible. Transitions of this kind occur in semiconductor systems with two phases having different chemical components (atoms) [71, 72]. In the situation under consideration, the phases are chemically identical, but have different structures and, what is particularly important for the spinodal-decomposition concept to be applicable, different vacancy concentrations.

It was assumed in [69] that vacancies are generated in the course of epitaxy at a constant rate v_{gen}, so that the concentration of carbon vacancies in the transition layer, $N_C(z)$ grows linearly as shown in Fig. 4.8a (here the z axis is directed along the hexagonal axis c, a = $z_2 - z_1$. This vacancy distribution was considered unstable. Therefore, it was assumed that spinodal decomposition occurs in the transition layer, and the vacancy distribution pattern takes the form shown in Fig. 4.8b.

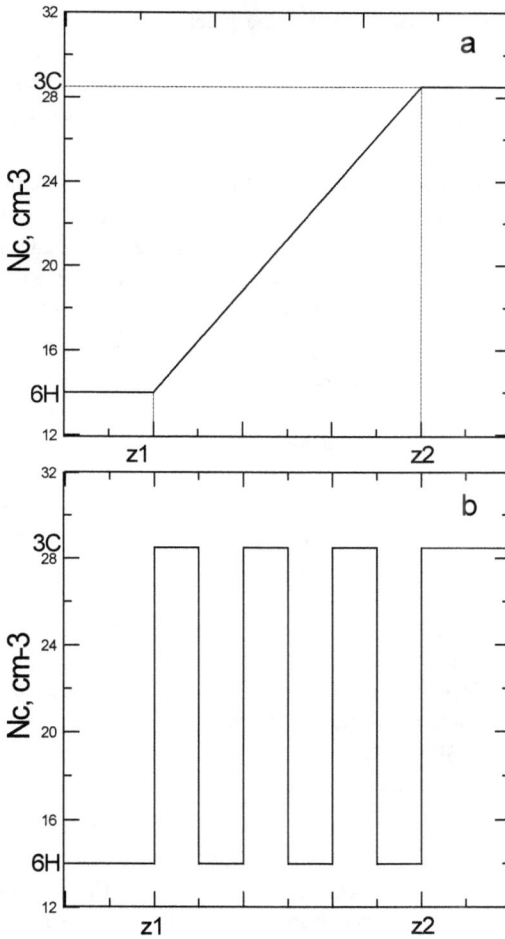

Fig. 4.8. Variation of the concentration of carbon vacancies along the c[0001] axis, corresponding to (a) unstable and (b) stable states of the system. $(z_2 - z_1)$ is the transition layer between the 6H and 3C polytypes [68].

A conclusion was made in [69] that the 6H polytype must occupy half of the transition layer when an epitaxial 3C layer is grown on a 6H substrate, with this result being a consequence of the linear run of $N_C(z)$ and valid in terms of the given (strongly simplified) model for the transition layer between any two polytypes.

On the whole, it can be concluded that the heteropolytype epitaxy of SiC is a rather unstable process. The problem of controlled growth of one polytype on the substrate of another polytype with an abrupt heterointerface is presently far from being fully solved. Therefore, a study of the nonepitaxial methods for fabrication of SiC-based heteropolytype structures is topical.

The degradation of the emission of blue SiC light-emitting diodes in the course of time was discovered as far back as 1981 [73]. An emission peak appeared in the green spectral range in the emission spectrum, and this emission was also spatially localized on the working plane of the diode. The authors of [73] explained their results by the formation of 3C-SiC inclusions in 6H-SiC p-n structures under the action of the forward current.

Later, after high-power rectifying diodes were fabricated from SiC, it was found that their characteristics deteriorate in the course of operation. This was manifested in an increase in the leakage current and decrease in the breakdown voltage under a reverse bias and in a rise in the diode resistance under a forward bias [74, 75]. This phenomenon has been studied extensively and it was found that the degradation is caused by the so-called stacking faults (SF), i.e., by the formation of interlayers of cubic SiC within 4H or 6H diodes through which a forward current is flowing [76 - 80 and references therein]. It was found that the energy released in the nonradiative recombination of an electron - hole pair in SiC is sufficient for overcoming the barrier for a shift of an atom to another position (9 meV/pair for 4H- and 4 meV/pair for 6H-SiC) [76]. As a result, the lattice of the hexagonal polytype was rearranged, with the atom shifted from hexagonal to cubic positions and an interlayer of the cubic polytype formed. Dislocations present in the bulk of the epitaxial layer or at the interface between this layer and the substrate served as starting points of the SF growth. Under the action of a forward current, the linear dimensions of the SF rapidly increased in the direction parallel to the c-axis of the crystal.

Commonly, this phenomenon is regarded as being adverse, because SF formation leads to deterioration of parameters power devices based on silicon carbide. At the same time, it has been shown that SFs with thicknesses of about several lattice constants constitute a 3C-SiC QW within a wider bandgap 4H-SiC. Such a structure caused quantum confinement and gave rise to a high-intensity photoluminescence (PL) in the blue spectral range (hν = 2.5 eV) [81].

It has also been found that SF quantum structures appear in heavily doped n-4H-SiC layers (N_D - N_A ~ 10^{19} cm^{-3}) after an additional thermal treatment. It was noted in [82-87] that thermally generated SFs would likely occur only if nucleation sites for partial dislocations existed in a crystal. Irrespective of the doping level, the residual stress in commercial 4H-SiC epilayers is large enough for nucleation and initiation of motion of partial dislocations at temperatures above 1000°C. It was concluded that internal stresses play a more important part in the nucleation of SFs than the doping level of a crystal. At the same time, it has been demonstrated that oxidation of lightly doped n-SiC layers does not lead to SF formation [88].

In [89], n- and p-type 4H-SiC crystals with various doping levels were subjected to thermal treatment. A transformation of the polytype was only observed in n-type samples with $N_D - N_A > 4 \times 10^{17}$ cm^{-3}, and not in p-type SiC. These results are in a good agreement with the concepts [46, 48] concerning the effect of impurities on the transformation of SiC polytypes. According to [46, 48], nitrogen (typical impurity for n-type SiC) favors the 6H $\Rightarrow$ 3C transformation, and introduction of aluminum (typical impurity for p-type SiC) promotes the 6H $\Rightarrow$ 4H transformation, i.e., it hinders the formation of the 3C polytype.

It was noted in [90 -92] that annealing of amorphized layers of hexagonal SiC polytypes leads not only to the recrystallization of the initial polytype, but also to the formation of inclusions of the cubic polytype 3C-SiC.

In our opinion, these results are rather important because the physics of mutual transformations of SiC polytypes is largely unclear. Of particular importance is the understanding of the reasons for a change of the polytype in an already grown epitaxial layer or a device structure. Unfortunately, the process in which SFs and the related QWs are formed in SiC remains uncontrollable.

Already before the studies of SFs by analysis of the structure and properties of deep centers in silicon carbide were commenced, a conclusion was made that there exists a concentration of native defects that is characteristic of each polytype [46, 58]. Thus, a change in the concentration of native defects may lead to a transformation of the polytype. It was suggested that a possible way to change the defect concentration in an already grown structure is by irradiation (possibly to the point of amorphization) with a subsequent annealing. The irradiation can create an increased concentration of defects both locally over an area (because of the small beam area) and locally within a volume (by using particles with different energies).

The development of such a technology could strongly expand the applicability field of SiC and affect studies of other polytype compounds. However, development of such a method of heterostructure fabrication requires that extensive studies of irradiation and annealing modes and preliminary doping of the starting epitaxial layers should be performed to confirm the possibility of its existence.

A difficulty encountered in studies of this kind is that hydrogen or helium bubbles are formed at high implantation doses, with the subsequent exfoliation of the upper layer of the semiconductor in annealing [93, 94].

4.3 Comparison of the radiation hardnesses of silicon and silicon carbide

4.3.1 Dependence of the radiation hardness on the functional purpose of a device

The first studies concerned with radiation defects in silicon carbide, carried in the 1950s-1960s confirmed the high radiation hardness of this material [95]. It should be noted that the crystals examined in those years were heavily doped and had a high density of structural defects. As increasingly perfect and pure SiC samples could be obtained, their experimentally measured radiation hardness gradually decreased. There even appeared reports in which it was stated that SiC detectors not only fail to surpass silicon detectors in radiation hardness, but are even inferior to the latter in a number of parameters (see Section 4.1) [96 -102].

As already noted in Section 4.1, the device parameters are mostly degraded under irradiation due to the compensation of the material by radiation defects. That is the radiation hardness is largely determined by the RD introduction rate [carrier removal rate (η_e)]. Earlier, the problem of the radiation hardness of wide-bandgap semiconductors was considered in [103]. It was noted that at least two approaches can be used to evaluate this parameter, depending on the functional purpose of a device.

In the first case (for most of the existing semiconductor devices), the radiation hardness is indirectly affected by other parameters of a semiconductor. In the second case (these are for the most part various types of detectors), only the carrier removal rates in both materials will be compared.

To illustrate the first approach, let us consider two diodes with the same breakdown voltage, those of silicon and silicon carbide types.

$$UbrSi = UbrSiC => (Ecr^{Si}\ W^{Si})/2 = (Ecr^{SiC}\ W^{SiC})/2 => W^{Si} = W^{Si} \times Ecr^{SiC}/ Ecr^{Si} \quad (4.3)$$

Here, Ubr is the breakdown voltage, Ecr is the critical electric field strength, and W is the thickness of the space-charge layer at Ubr.

Taking into account that $Ecr^{SiC}/ Ecr^{Si} \sim 10$, a $W \sim (Nb)^{1/2}$, where Nb is the uncompensated impurity concentration in the base region of the diode, we have

$$Nb^{SI} = 100\ Nb^{SiC} \quad (4.4)$$

This means that, at the same breakdown voltage, the base of the SiC diode will be doped to a two orders of magnitude higher level, compared with the base of the Si diode. Consequently, even at equal values of η_e, the compensation of a silicon carbide diode will require a 100 times larger irradiation dose, compared with the compensation of a Si diode.

In the second case, typical of detectors, the reverse voltage applied to a device is limited to several hundred volts. It is required that the base layer should be as thick as possible. Because $W \sim (Nb)^{1/2}$, then, comparing the silicon- and silicon carbide-based detectors, we have:

$$W^{Si} \sim W^{SiC} => Nb^{Si} \sim Nb^{SiC}$$

That is the comparison of the radiation hardnesses reduces to comparison of the values of η_e. Although $\eta_e^{Si} \sim 2\,\eta_e^{SiC}$, the gain in radiation hardness will be substantially lower than that for high-power devices.

4.3.2 Effect of temperature on the radiation hardness

It was shown in Section 2.2 that, in contrast to narrow-gap semiconductors, deep centers that are not ionized at room temperature can be formed in wide-bandgap semiconductors, and the contribution of these centers to η_e may be dependent on the temperature at which a measurement is made [104]. That is the experimentally measured value of η_e will decrease with increasing measurement temperature.

Further, we would like to consider another aspect of η_e determination in wide-bandgap semiconductors, specifically, the effect of elimination of radiation defects by annealing. It is known that the interaction of charged particles with the crystal lattice of the semiconductor yields vacancies and interstitials, the so-called primary radiation defects. If the semiconductor is irradiated at low temperatures (~ 0 K), these primary defects hardly recombine, being in the "frozen" state. As the temperature increases, these defects start to move over the crystal, with most of these defects recombining and the rest forming complexes with impurity atoms or divacancies, i.e., more thermally stable secondary radiation defects. And finally, as the temperature increases even further, the secondary defects are also annealed and the crystal returns to its initial (unirradiated) state.

Table 4.1. Annealing temperatures of radiation defects in SiC upon various kinds of irradiation

Refs.	SiC	Type of irradiation	Transformation onset temperature of defects, °C	Final annealing temperatures of RDs, °C
[105]	4H-p	e-2,5 MeV	200-400	950-1400
[106]	6H-n	e- 0,3-0,4 MeV	400-900	1600
[107]	4H- n	e-15 MeV; p-1,2 MeV	200 -800	> 1200
[108]	6H-n 4H- n	e-2,5 MeV; p-1 MeV, He		1200-1700
[109]	4H- n	e-15 MeV	400-800	1200-2000

Here, e designates the irradiation with electrons; p, that with protons; and He, that with helium nuclei.

The stages of annealing of radiation defects have been well studied for silicon. It is known that primary defects start to be annealed at a temperature of ~80 K, and secondary defects (A and E centers, divacancies) are annealed at temperatures of 150-350 °C [104].

Recently, a number of studies of how radiation defects are annealed in silicon carbide have been carried out. Their results are partly presented in Table 4.1.

It can be seen in the table that two stages of annealing can be also distinguished for silicon carbide. In the first of these at 200 – 900 °C, DLTS measurements demonstrated a decrease in the concentrations of some radiation defects and an increase in the concentration of others. In the second stage at temperatures exceeding 1000 – 1200 °C, the radiation defects were completely eliminated by annealing.

Figure 4.9 shows schematically the variation of the concentration of radiation defects formed under the same irradiation of Si and SiC at low temperatures (conditionally 0 K). It can be seen in the figure that Si and SiC are strongly different at room temperature as regards the spectra of the radiation defects existing in these semiconductors. The annealing of primary defects has already ended in silicon, but is not yet commenced in SiC. As a result, η_e measured at 300 K for SiC may be found to be the same as that in Si or even higher. Possibly, measurements of this kind will be of practical interest, but it is incorrect to make conclusions about the radiation hardness of SiC on the basis of these data.

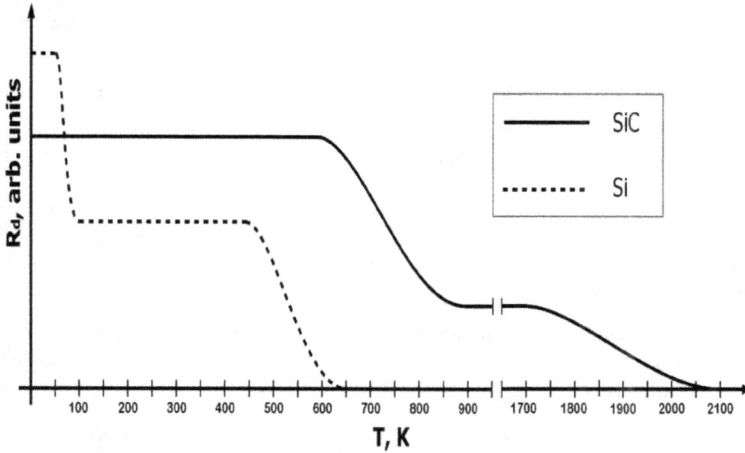

Fig. 4.9. Schematic representation of the annealing of radiation defects in silicon and silicon carbide [107].

In our opinion, it is possible to compare η_e for Si and SiC either for values measured at low temperature (0 K) before the beginning of the stage in which primary defects are annealed, or upon completion of this stage, which corresponds to 200-300 K for Si and 800—900 °C for SiC.

From the standpoint of semiconductor electronics, this means that the highest radiation hardness can be obtained for SiC devices if these devices are used at elevated temperatures. Possibly, it is possible to envisage the annealing of a SiC device directly in the working circuit due to its self-heating when high-power current pulses are passed.

4.4   Conclusion

In this book, we briefly considered the main results obtained in studies of the effect of charged particles on the properties of silicon carbide and parameters of devices based on this material. When writing the book, we tried to focus on the specific features of radiation defects in SiC as a wide-bandgap semiconductor. We also tried to make a comparison of silicon and SiC.

It can be concluded that many aspects of the radiation physics of silicon carbide have been studied in sufficient detail. Among the most interesting results, the following can be singled out.

Various kinds of irradiation result in that carbon vacancies V_C are formed in the crystal lattice of SiC, and these vacancies are stable up to high annealing temperatures. To these RDs are related deep centers (R in 6H-SiC and $Z_{1/2}$ in 6H-SiC), which are main nonradiative recombination centers.

The carrier removal rate in SiC and, apparently, in other wide-bandgap semiconductor depends on the measurement temperature, and at low temperatures, also on the method used to determine the carrier concentration.

The compensation of silicon carbide under irradiation occurs because carriers pass to the RDs being formed. As a result, a linear decrease in Nd - Na (Na - Nd) is observed with increasing irradiation dose.

The irradiation results in that silicon vacancies are formed in SiC, which are similar in their properties to NV centers in diamond and show certain promise for further application in spintronics.

It was shown that the main influencing factor is the energy of primary knock-on atoms (PKAs). As this energy increases, first, the average distance between the genetically related FPs becomes larger, and, as a consequence, the fraction of FP unrecombined under irradiation grows. Second, new, more complex secondary radiation defects can be formed as the PKA energy increases.

The photoluminescence characteristic of SiC, which appears upon irradiation with various kinds of charged particles (DPL), is in all probability due to the radiative recombination via the nitrogen - radiation defect (V_C ?) donor - acceptor pair.

The radiation hardness of SiC devices (except detectors) is approximately two orders of magnitude higher than that of silicon-based devices. The radiation hardness of SiC devices must grow with increasing irradiation temperature.

Despite the success achieved in studies of radiation defects in SiC, there exist a number of areas that, in our opinion, invite further examination.

There hardly have been reports on studies of SiC irradiated at elevated temperatures.

It is unclear whether it is possible to produce controlled transformation of the silicon carbide polytype under irradiation.

The formation and properties of radiation defects in 3C SiC have been insufficiently studied, which is possibly due to the poor structural perfection of the existing material.

Problem of the influence exerted by the nonequivalent positions in the SiC lattice on the ionization energies of impurity and defect centers has not been studied to the full extent so far.

4.5 Acknowledgments

The author considers his pleasant duty to thank V. V. Kozlovskii, professor of Peter the Great St. Petersburg Polytechnic University, for fruitful discussions in writing all the sections of this book. The author is also grateful to staff members of the Ioffe Physical-Technical Institute A. M. Strel'chuk, N. B. Strokan, A. M. Ivanov, D. V. Davydov, S. Yu. Davydov, N. V. Seredova, S. P. Lebedev, E. V. Kalinina, E. V. Bogdanova, and I. P. Nikitina and to a staff member of Tallinn University of Technology O. M. Korol'kov for assistance in experiments and helpful remarks. The study was in part supported by the Russian Science Foundation (project no. 16-12-10106 "Radiation hardness of silicon carbide and devices based on this material for extreme electronics").

References

[1]   A. V. Afanasev, V. A. Ilin, I. G. Kazarin, A. A. Petrov, Thermal stability and radiation hardness of SiC-based schottky-barrier diodes, Technical Physics 46, pp 584-586, (2001) http://dx.doi.org/10.1134/1.1372950

[2]   G.Izzo, G.Litrico, L.Calcagno, G.Foti and F.La Via, Electrical properties of high energy ion irradiated 4H-SiC Schottky diode, J.Appl.Phys. 104, 093711 (2008). http://dx.doi.org/10.1063/1.3018456

[3]   E.Omotso, W.E.Meyer, F.D.Auret, A.T.Paradzah, M.Diale, S.M.M. Coelho, P.J.Janse van Rensburg, The influence of high energy electron irradiation on the Schottky barrier height and Richardson constant of Ni/4H-SiC Schottky diodes, Materials Science in Semiconductor Processing, 39 (2015) 112-118. http://dx.doi.org/10.1016/j.mssp.2015.04.031

[4]   A.A. Lebedev, K. S. Davydovskaya, A. M. Strelchuk, V. V. Kozlovsky, Radiation resistance of 4H-SiC Schottky diodes with electron irradiation with energy of 0.9 MeV, To be published in Journal of Surface Investigation: X-ray, Synchrotron and Neutron Techniques (2017).

[5]   V. V. Emtsev, A. M. Ivanov, V. V. Kozlovski, A. A. Lebedev, G. A. Oganesyan, N. B. Strokan, G. Wagner, Similarities and distinctions of defect production by fast electron and proton irradiation: Moderately doped silicon and silicon carbide of n-type, Semiconductors, 2012, Vol. 46, No. 4, pp. 456-465 http://dx.doi.org/10.1134/S1063782612040069

[6]   V. V. Kozlovski, Lebedev, A. M. Strelchuk, K. S. Davidovskaya, A. E. Vasilev, L. F. Makarenko, Effect of the bombarding electrons energy on epitaxial layer n-4H-SiC (CVD) conductivity, accepted for publication in Semiconductors (2017)

[7]   A.A. Lebedev, K.S. Davydovskaya, V.V. Kozlovski, O.M. Korolkov, N.Sleptsuk, J.Toompuu Degradation of 600-V 4H-SiC Schottky diodes under irradiation with 0.9 MeV electrons, to be published in Material.science Forum (2016).

[8]   A. Sciuto, F. Roccaforte and V. Raineri, Electro-optical response of ion-implanted 4H-SiC Schottky ultraviolet photodetectors, Appl.Phys.Lett 92, 093505 (2008) http://dx.doi.org/10.1063/1.2891048

[9]   A.Cavallini, A.Castaldini and F.Nava, On the UV responsity of neutron irradiated 4H-SiC, Appl.Phys.Lett. 93, 153502 (2008) http://dx.doi.org/10.1063/1.2993224

[10]   E. V. Kalinina, A. A. Lebedev, E. Bogdanova, B. Berenquier, L. Ottaviani,G. N. Violina, V. A. Skuratov, Irradiation of 4H-SiC UV detectors with heavy ions, Semiconductors, 49 (2015) 540-546. http://dx.doi.org/10.1134/S1063782615040132

[11]   C.Canali, M.Martini, G.Ottaviani, A.Alberigi Quaranta, Measurements of the average energy per electron-hole pair generation in silicon between 5-320 K IEEE Trans NS-19 (4) 9 (1972).

[12]   R.D.Ryan, Precision Measurements of the Ionization Energy and Its Temperature Variation in High Purity Silicon Radiation Detectors Nucl.Instr. Meth, 120(1) 201 (1974) http://dx.doi.org/10.1016/0029-554X(74)90128-1

[13]   A. A. Lebedev, N. S. Savkina, A. M. Ivanov, N. B. Strokan, D. V. Davydov, 6H-SiC epilayers as nuclear particle detectors, Semiconductors, 34 (2000) 243-249. http://dx.doi.org/10.1134/1.1187940

[14]   F.H.Ruddy, A.R.Dulloo, J.G.Seidel, S.Seshadri, and L.B.Rowland, Developing of a Silicon Carbide Radiator Detector, IEEE Transaction on nuclear Science, 45 (1998) 536- 541. http://dx.doi.org/10.1109/23.682444

[15]   A.R.Dulloo, F.H.Ruddy, J.G.Seidel, C. Davison, T.Flinchbaugh and Daubenspeck, Simultaneous Measurement of Neutron and Gamma-Ray Radiation Levels from a TRIGA Reactor Core Using Silicon Carbide Semiconductor Detectors, IEEE Transaction on nuclear Science, 46 (1999) 275- 279 http://dx.doi.org/10.1109/23.775527

[16] M.Bruzzi, F.Nava, S.Russo, S. Sciortino, P. Vanni, Characterization of silicon carbide detectors response to electron and photon irradiation, Diamond and Rel mat., 10 (2001) 657-661.

[17] G.Bertuccio, R.Casiraghi and F.Nava, Epitaxial Silicon Carbide for X-ray Detection, IEEE Transaction on nuclear Science, 48 (2001) 232- 233 http://dx.doi.org/10.1109/23.915369

[18] R.Stevenson, Silicon carbide opens the door to radiation-detection market, Compound Semiconductors, January/February 2005, 29-30.

[19] R.Schifano, A.Vinattieri, M.Bruzzi, S.Miglio, S.Lagomarsino, S.Sciortino, F.Nava, Electrical and optical characterization of 4H-SiC diodes for particle detection, J.Appl.Phys 97, 103539 (2005). http://dx.doi.org/10.1063/1.1906294

[20] A.M.Ivanov, E.V. Kalinina, A. O. Konstantinov, G. A. Onushkin, N. B. Strokan, G. F. Kholuyanov, A. Hallén, High-resolution short range ion detectors based on 4H-SiC films Technical Physics Letters 2004, V 30, pp 575-577 http://dx.doi.org/10.1134/1.1783406

[21] A. M. Ivanov, M. G. Mynbaeva, A. V. Sadokhin, N. B. Strokan, A. A. Lebedev, Charge transport in 4H-SiC detector structures under conditions of a high electric field, Semiconductors, 43 (2009) 1052-1054 http://dx.doi.org/10.1134/S1063782609080168

[22] A. M. Ivanov, N. B. Strokan,V. V. Kozlovskii, A. A. Lebedev, Effect of electron and proton irradiation on characteristics of SiC surface-barrier detectors of nuclear radiation, Semiconductors, 42 (2008) 363-369. http://dx.doi.org/10.1134/S1063782608030238

[23] N.Iwamoto, B.C. Johnson, N. Hosino, M.Ito, H.Tsuchida, K. Kojima, Defect - induced performance degradation of 4H-SiC Schottky barrier diodes, J.Appl.Phys 113, 143714 (2013) http://dx.doi.org/10.1063/1.4801797

[24] Z.Luo, T.Chen, J.D.Cressler, D.C.Sheridan, J.R.Williams, R.A.Reed and P.W.Marshall, Impact of Proton Irradiation on the Static and Dynamic Characteristics of High-Voltage 4H-SiC JBS Switching Diodes, IEEE trans on Nucl.Sci 50 (2003) 1821-1826. http://dx.doi.org/10.1109/TNS.2003.821806

[25] G.Alferi, A.Mihaila, H.M.Ayedh, B.G.Svensson, P.Hazdra, P.Godignon, J.Milan, and S.Kicin, Deep Level Characterization of 5MeV proton irradiated SiC PiN diodes, Mater.Sci.Forum 858 (2016) 308-322. http://dx.doi.org/10.4028/www.scientific.net/MSF.858.308

[26]   G. N. Violina, E. V. Kalinina, G. F. Kholujanov, G. A. Onushkin, V. G. Kossov, R. R. Yafaev, A. Hallén, A. O. Konstantinov, Photoelectric properties of p+-n junctions based on 4H-SiC ion-implanted with aluminum, Semiconductors, 36, pp 706-709 (2002) http://dx.doi.org/10.1134/1.1485675

[27]   A. M. Ivanov, E. V. Kalinina, N. B. Strokan, Charge transfer in the presence of a layer of trapping centers in semiconductor SiC ionizing radiation detectors, Technical Physics Letters, 34, p1069 (2008) http://dx.doi.org/10.1134/S1063785008120249

[28]   E. V. Kalinina, N. B. Strokan, A. M. Ivanov, A. A. Sitnikova, A. V. Sadokhin, A. Yu. Azarov, V. G. Kossov, and R. R. Yafaev, High-Temperature Nuclear - Detector Arrayes Based on 4H-SiC Ion-Implantation-Doped p+-n Junction, Semiconductors, 42, (2008) 86-91. http://dx.doi.org/10.1134/S1063782608010120

[29]   A. M. Ivanov, N. B. Strokan, A. A. Lebedev, Correction of the characteristics of strongly irradiated SiC-based nuclear radiation detectors by increasing the working temperature,Technical Physics Letters, January 2009, Volume 35, pp 50-53 (2009)

[30]   A. M. Ivanov, N. B. Strokan, E. V. Bogdanova, A. A. Lebedev, Instability of characteristics of SiC detectors subjected to extreme fluence of nuclear particles, Semiconductors, 41, pp 115-119 (2007) http://dx.doi.org/10.1134/S1063782607010216

[31]   http://cree.com/

[32]   N.Iwamoto, S.Onoda, S.Hishiki, T.Ohshima, M.Murakami, I.Nakano, and K. Kawano, Degradation of Charge Collection Efficiency for 6H-SiC Diodes by Electron Irradiation, Materials Science Forum, 600-603, pp1043-1046 (2009) http://dx.doi.org/10.4028/www.scientific.net/MSF.600-603.1043

[33]   V. V. Emtsev, A. M. Ivanov, V. V. Kozlovski, A. A. Lebedev, G. A. Oganesyan, N. B. Strokan, G. Wagner, Similarities and distinctions of defect production by fast electron and proton irradiation: Moderately doped silicon and silicon carbide of n-type, Semiconductors, 46, pp 456-465 (2012) http://dx.doi.org/10.1134/S1063782612040069

[34]   G. Casse, Overview of the recent activities of the RD50 collaborationon radiation hardening of semiconductor detectors for the sLHC, Nuclear Instruments and Methods in Physics Research A 598 (2009) 54-60 http://dx.doi.org/10.1016/j.nima.2008.08.019

[35]   J. Metcalfe, Silicon Detectors for the sLHC, Nuclear Physics B (Proc. Suppl.) 215
        (2011) 151-153 http://dx.doi.org/10.1016/j.nuclphysbps.2011.03.162

[36]   J.N.Merrett, J.R.Williams, J.D. Cressler, A.Sutton, L.Cheng, V.Bondarenko,
        I.Sankin, D.Seale, M.S.Mazzola, B.Krisnan, Ya.Koshka, and J.B.Casady, Gamma
        and proton Irradiation Effects on 4H-SiC Depletion-Mode Trench JFETs,
        Mater.Sci.Forum, 483-485 (2005) 885-888.
        http://dx.doi.org/10.4028/www.scientific.net/MSF.483-485.885

[37]   Zhang Lin, Zhang Yimen, Zhang Yuming, and Han Chao, Neutron radiation effect
        on 4H-SiC MESFETs and SBDs, Journal of Semiconductors, V. 31, (2010)
        114006. http://dx.doi.org/10.1088/1674-4926/31/11/114006

[38]   M.Yoshikawa, H.Itoh, Y.Morita, I.Nashiyama, S.Missawa, H.Okumura, and
        S.Yoshida, Effects of gamma-ray irradiation on cubic silicon carbide metal-oxide-
        semiconductor structure, J.Appl.Phys, 70 (1991) 1309-1312.
        http://dx.doi.org/10.1063/1.349587

[39]   M.Yoshikawa, K. Saitoh, T.Ohshima, H.Itoh, Y.Morita, I.Nashiyama, Y.
        Takahashi, K. Ohnishi, H.Okumura, and S.Yoshida, Generation Mechanism of
        Trapped Charges in Oxide Layers of 6H-SiC MOS Structures Irradiated with
        Gamma-Rayes, Mater.Sci.Forum 264-268 (1998) 1017-1020.
        http://dx.doi.org/10.4028/www.scientific.net/MSF.264-268.1017

[40]   K. Ohnishi, H.Itoh, and M.Yoshikawa, Effect of gamma-ray irradiation on the
        characteristics of 6H-SiC silicon carbide metal-oxide-semiconductor field effect
        transistor with hydrogen-annealed gate oxide, J.Appl.Phys., V 90, (2001) 3038-
        3041. http://dx.doi.org/10.1063/1.1394895

[41]   S.Mitra, M.V.Rao, K.Jones, N.Papanicolaou, and S.Wilson, Deep levels in ion
        implanted field effect transistors on SiC, Solid-state Electronics, 47 (2003) 193-
        198. http://dx.doi.org/10.1016/S0038-1101(02)00194-6

[42]   C.Claeys, E. Simoen, Radiation Effects in Advanced Semiconductor Materials and
        Devices, Berlin, Heidelberg, New York, Barcelona, Hongkong, Milan, Paris,
        Tokyo: Springer, 2002. http://dx.doi.org/10.1007/978-3-662-04974-7

[43]   V.M.Polyakov and F.Schwierz , Calculation of Cutoff frequency fT for a Model
        6H-SiC/3C-SiC Field-effect Transistor, was presented at 48 Intern.
        Wissenschaftliches colloquium Technische Universitat Ilmenau, 22-25 sept 2003.

[44] V.M.Polyakov and F.Schwierz, Formation of two-dimentional electron gases in polytypic Sic heterostructures, J.Appl.Phys. 98 (2005) 023709. http://dx.doi.org/10.1063/1.1984070

[45] C.C. Chen, A.B.Horsfall, N.G.Wrigth, A.G.O. Neill Optimization of heterostructure Bipolar Transistors in SiC. Materials Science Forum Vols. 483-485(2005) 913. http://dx.doi.org/10.4028/www.scientific.net/MSF.483-485.913

[46] Yu.A.Vodakov, G.A.Lomakina and E.N.Mokhov, Non-stoichiometry and Polytypism of Silicon carbide. Sov.Phys.Solid State 24 (1982) 780.

[47] Yu.Vakhner and Yu.M.Tairov, About SiC (Sc) polytypism, grown from melt. Sov.Phys.Solid State 12 (1970) 1213.

[48] Yu.A.Vodakov, E.N.Mokhov, A.D.Roenkov and M.M.Anikin Impurity effects on Silicon Carbide Polytypism, Sov.Tech.Phys.Lett. 5 (1979) 147.

[49] A.A.Maltsev, A.Yu.Maksimov, and N.K.Yushin, 4H-SiC single crystal ingots grown on 6H-siC and 15R-SiC seeds. Inst.Phys.Conf.Ser. 142, 42 (1996).

[50] A.A.Maltsev, D.P.Litvin, M.P.Scheglov, and I.P.Nikitina, high-resistivity epitaxial films of 4H-SiC doped by scandium, Inst.Phys.Conf.Ser.142,137 (1996)

[51] P.A.Ivanov,A.A.Maltsev,V.N.Panteleev, T.P.Samsonova, A.Yu.Maksimov, N.K.Yushin, and V.E.Chelnokov, 4H-SiC field-effect transistor hetero-epitaxially grown on 6H-SiC substrate by sublimation. Inst.Phys.Conf.Ser.142,757 (1996).

[52] A.A.Lebedev and N.A.Sobolev, Capacitance spectroscopy of deep centers in Sic, Material.science Forum 258-263 (1997) 715.

[53] F.C.Frank, The growth of carborundum; dislocation and polytypism, Phil.Mag. 42 (1951) 1014. http://dx.doi.org/10.1080/14786445108561346

[54] V.Vand, J.I.Hanoka, Epitaxial theory of polytypism; observations on the growth of PbI2 crystals, Mat.Res.Bull 2 (1967) 241. http://dx.doi.org/10.1016/0025-5408(67)90063-3

[55] S.Mardix, I.T.Steinberg Allotropic transformations in ZnS Isr.J.Chem 3 (1966) 243.

[56] A.Fissel, Artificially layered heteropolytypic structures based on SiC polytypes: molecular beam epitaxy, characterization and properties. Phys.Rep 379(2003) 149. http://dx.doi.org/10.1016/S0370-1573(02)00632-4

[57]   N. D. Sorokin, Yu. M. Tairov, V. F. Tsvetkov, and M. A. Chernov, Investigation of the crystal-chemical properties of the silicon carbide polytypes, Kristallografiya 28, 910-914 (1983).

[58]   A.A.Lebedev Influence of native defects on polytypism in SiC, Semiconductors 33 (1999) 107. http://dx.doi.org/10.1134/1.1187657

[59]   A.A.Lebedev and S.Yu.Davydov A vacancy model of the heteropolytype epitaxy process. Semiconductors 39 (2005) 277. http://dx.doi.org/10.1134/1.1882785

[60]   Chien F R, Nutt S R, Yoo W S, Kimoto T and Matsunami H Terrace growth and polytype development in epitaxial ß-SiC films on a-SiC (6H and 15R) substrates J. Mater.Res. 9, 940 (1994) http://dx.doi.org/10.1557/JMR.1994.0940

[61]   Dudley M, Vetter W M and Neudeck P G, Polytype identification in heteroepitaxial 3C-SiC grown on 4H-SiC mesas using synchrotron white beam x-ray topography, J. Cryst. Growth 240, 22 (2002) http://dx.doi.org/10.1016/S0022-0248(02)00827-8

[62]   Rowland L B, Kern R S, Tanaka S and Davis R F, Gas-source molecular beam epitaxy of monocrystalline ß-SiC on vicinal a(6H)-SiC, J. Mater. Res. 8, 2753 (1993) http://dx.doi.org/10.1557/JMR.1993.2753

[63]   Lebedev A A, Savkina N S, Strelchuk A M, Tregubova A S and Scheglov M P 6H-3C SiC structures grown by sublimation epitaxy Mater. Sci. Eng. B 46, 168, (1997) http://dx.doi.org/10.1016/S0921-5107(96)01956-3

[64]   Lebedev A A, Strelchuk A M, Davydov D V, Savkina N S, Tregubova A S, Kuznetsov A N, Soloviev V A and Poletaev N K, Electrical characteristics of p-3C-SiC/n-6H-SiC heterojunctions grown by sublimation epitaxy on 6H-SiC substrates, Appl. Surf. Sci. 184, 419 (2001) http://dx.doi.org/10.1016/S0169-4332(01)00530-X

[65]   D. B. Shustov, A. A. Lebedev, S. P. Lebedev, D. K. Nelson, A. A. Sitnikova, and M. V. Zamoryanskaya, Comparative Study of 3C-SiC Layers Sublimation_Grown on a 6H_SiC Substrate, Semiconductors, 2013, Vol. 47, No. 9, pp. 1267-1270. http://dx.doi.org/10.1134/S1063782613090236

[66]   R. Vasiliauskas, M. Syväjärvi, M. Beskova, R. Yakimova, Mater. Two - dimensional Nucleation of Cubic and 6H Silicon Carbide, Sci. Forum 615 - 617 (2009) 189. http://dx.doi.org/10.4028/www.scientific.net/MSF.615-617.189

[67]   R. Vasiliauskas, M. Marinova, M. Syväjärvi, A. Mantzari, A. Andreadou, J. Lorenzzi, G. Ferro, E. K. Polychroniadis, R. Yakimova, Sublimation Growth and

Structural Characterization of 3C-SiC on Hexagonal and Cubic SiC Seeds, Mater.Sci.Forum 645- 648 (2010) 175.
http://dx.doi.org/10.4028/www.scientific.net/MSF.645-648.175

[68] A.A.Lebedev, M.V.Zamoryanskaya, S.Yu.Davydov, D.A.Kirilenko, S.P.Lebedev, L.M. Sorokin, D. B.Shustov, M.P.Shcheglov, A study of the intermediate layer in 3C-SiC/6H-SiC heterostructures, Journal of Crystal Growth 396 (2014)100-103
http://dx.doi.org/10.1016/j.jcrysgro.2014.03.030

[69] A.A. Lebedev, M.V. Zamoryanskaya, S. Yu. Davydov, D.A. Kirilenko, S.P. Lebedev, L.M. Sorokin, D.B. Shustov, and M.P. Shcheglov, Investigation of the Transition Layer in 3C_SiC/6H_SiC Heterostructures, Semiconductors, 2013, Vol. 47, No. 11, pp. 1539-1543. http://dx.doi.org/10.1134/S1063782613110134

[70] V.P. Skripov, A.V. Skripov., Spinodal decomposition (phase transitions via unstable states), Sov. Phys. Usp. 22 p.389-410 (1979)
http://dx.doi.org/10.1070/PU1979v022n06ABEH005571

[71] N.A. Bert, L.S. Vavilova, I.P. Ipatova, V.A. Kapitonov, A.V. Murashova, N.A. Pikhtin, A.A. Sitnikova, I.S. Tarasov, V.A. Shchukin. Spontaneously forming periodic composition-modulated InGaAsP structures, Semiconductors, 33, pp 510-513, (1999) http://dx.doi.org/10.1134/1.1187719

[72] A.Yu. Maslov, O.V. Proshin. Effect of the spectrum of elementary excitations on spinodal decomposition of semiconductor alloys, Semiconductors, Volume 43, pp 841-845, (2009) http://dx.doi.org/10.1134/S1063782609070033

[73] G.Zeither, D. Theeis.A New degradation Phenomenon in Blue light Emitting Silicon carbide Diodes. IEEE Trans.Electron.dev. ED-28, (1981) 425.

[74] J.P.Bergman, H.Lendenmann, P.A.Nilsson, U.Lindefelt and P.Skytt, Crystall defects as Source of Anomalous Forward Voltage Increase of 4H-SiC Diodes. Mat.Science Forum 353-356 (2001) 299.
http://dx.doi.org/10.4028/www.scientific.net/MSF.353-356.299

[75] H.Lendenmann, F.Dahlquist, N.Johansson, R.Soderholm, P.A.Nilsson, J.P.Bergman and P.Skytt, Long Term Operation of 4,5 kV PiN and 2,5kV JBS Diodes. Mater.Science Forum 353-356 (2001) 727.
http://dx.doi.org/10.4028/www.scientific.net/MSF.353-356.727

[76] M.S.Mao, S.Limpijumnong and W.R.L.Lambrecht, Stacking faults band structure in 4H-SiC and its impact on electron devices. Appl.Phys.Lett. 79(26) (2001) 4360.
http://dx.doi.org/10.1063/1.1427749

[77] H.Iwata, U.Lindefelt, S.Oberg, P.R.Briddon, Cubic polytype inclusions in 4H-SiC. J.Appl.Phys. 93(3) (2003) 1577. http://dx.doi.org/10.1063/1.1534376

[78] K.Kojima,T.Ohno, T.Fujimo, M.Katsumo, N.Ohtani, J.Nishio, T.Suzuki, T.Tanaka, Y.Ishida, T.Takahashi and K.Arai, Influence of stacking faults on the performance of 4H-SiC Schottky barrier diodes fabricated on (11-20)face. Appl.Phys.Lett. 81(16) 2002, 2974. http://dx.doi.org/10.1063/1.1512956

[79] M.E.Twigg, R.E.Stahlbush, M.Fatemi, S.D.Arthur, J.B.Fedison, J.B.Tucker and S.Wang, Structure of stacking faults formed during the forward bias of 4H-SiC p-i-n diodes. Appl.Phys.lett. 82(15) (2003) 2410. http://dx.doi.org/10.1063/1.1566794

[80] M.Skowronski and S.Ha, Degradation of hexagonal silicon-carbide-based bipolar devices. J.Appl.Phys. 99, 011101 (2006). http://dx.doi.org/10.1063/1.2159578

[81] R. Okojie, M.Xhang, P.Pirouz, S.Tumakha, G.Jessen. Observation of 4H-SiC to 3C-SiC polytypic transformation during oxidation, Appl.Phys.Lett. 79, 3056 (2001) http://dx.doi.org/10.1063/1.1415347

[82] L.J. Brillson, S.Tumakha, G.H.Jessen, R.S.Okojie, M.Zhang, P.Pirouz. Thermal and doping dependence of 4H-SiC polytype transformation Appl.Phys.Lett., 81, 2785 (2002). http://dx.doi.org/10.1063/1.1512816

[83] T.A.Kuhr,J.Q.Liu, H.J.Chung, M.Skowronski and F.Szmulowicz, Spontaneous formation of stacking faults in highly doped 4H-SiC during annealing. J.Appl.Phys. 92(10) 5863. http://dx.doi.org/10.1063/1.1516250

[84] T.Suzuki Suppressuion of oxidation-induced stacking fault generation in argon ambient annealing with controlled oxygen and the effect upon bulk defects. Mat.Science in Sem.Proc. 5 (2002) 5.

[85] J.Q.Liu, H.J.Chung, T.Kuhr, Q.Li and M.Skowronski, Structural instability of 4H-SiC polyttype induced by n-doping, Appl. Phys. Lett. 80 2111 (2002) http://dx.doi.org/10.1063/1.1463203

[86] R.S. Okojie, M.Xang, P.Pirouz, S.Tumakha, G.Jessen and L. Brillson, 4H- to 3C-SiC Polytypic Transformation during Oxidation Mat.Science Forum 389-393 (2002) 451. http://dx.doi.org/10.4028/www.scientific.net/MSF.389-393.451

[87] R.S.Okojie, M.Zhang and P.Pirouz, residual stress and stacking Faults in N-type 4H-SiC Epilayers Mat.Science Forum 457-460 (2004)529. http://dx.doi.org/10.4028/www.scientific.net/MSF.457-460.529

[88] V.V.Afanasev, F.Ciobanu, S.Dimitrijev, G.Pensl and A.Stesmans, SiC/SiO2 Interface State: Properties and Models, Mat.Science Forum 483-485 (2005)451.

[89] L.J.Brillson,S.Tumakha, G.H.Jessen,R.S.Okojie, M.Zhang and P.Pirouz. Thermal and doping dependence of 4H-SiC polytype transformation. Appl.Phys.Lett. 81(15) (2002) 2785. http://dx.doi.org/10.1063/1.1512816

[90] H.Inui, H.Mori and H.Fujita. Electron-irradiation-induced crystalline to amorphous transition in ?-SiC single crystals. Philosophical Magazine B, 1, 107 (1990). http://dx.doi.org/10.1080/13642819008208655

[91] A.A.Lepneva, E.N.Mokhov, V.G.Oding A.S.Tregubova, Silicon carbide irradiated by high doses of neutrons. Sov.Phys,Solid State, 33, 7- (1991) 221.

[92] P.Musumeci, L.Calcagno, M.G.Grimaldi, G.Foti. Optical defects in ion damaged 6H-silicon carbide Nuclear Instr. Meth. Phys. Research B, 116, (1996) 327.

[93] E. Oliviero, M.L.David, M.F.Beaufort, J.Nomgaudyte, L.Pranevicius, A.Declemy, J.F.Barbot. Formation of bubbles by high dose He implantation in 4H-SiC J.Appl.Phys, 91, 1179 (2002). http://dx.doi.org/10.1063/1.1429760

[94] E.V.Bogdanova, V.V.Kozlovski, D.S.Rumiantsev,A.A.Volkova, A.A.Lebedev, Modification of Silicon Carbide by proton irradiation Mat.Science.Forum 457-460 (2004) 249. http://dx.doi.org/10.4028/www.scientific.net/MSF.457-460.817

[95] W.J.Choyke. A review of radiation damage in SiC. Inst.Phys.Conf.Ser., 31, (1977) 58-69

[96] A.Hallen, A.Henry, P. Pelligrino, B.G.Swensson, D.Aberg, Ion implantation induced defects in epitaxial 4H-SiC, Material Science and Eng, B61-62 (1999) 378-381. http://dx.doi.org/10.1016/S0921-5107(98)00538-8

[97] B.G.Swensson, A. Hallen, M. K. Linnarson, A. Yu. Kuznetsov, M.S. Janson, D. Aberg, J. Osterman, P. O. A. Persson, L. Hultman, L. Storasta, F. H. Carlsson, J. P. Bergman, C. Jagadish, and E. Morvan, Doping of Silicon Carbide by ion Implantation, Material Science Forum, 353-356 (2001) 549 -554. http://dx.doi.org/10.4028/www.scientific.net/MSF.353-356.549

[98] G.Casse, Overview of the recent activities of the RD50 collaboration on radiation hardening of semiconductor detectors for the sLHC, Nuclear Instruments and Methods in Physics Reserch A 598 (2009) 54-60. http://dx.doi.org/10.1016/j.nima.2008.08.019

[99] J.Metcalfe, Silicon Detectors for the sLHC, Nuclear Physics B (Proc.Suppl.) 215 (2011) 151-153. http://dx.doi.org/10.1016/j.nuclphysbps.2011.03.162

[100] A. A. Lebedev, V.V.Kozlovski, N. B. Strokan, D.V.Davydov, A. M. Ivanov, A. M. Strel'chuk, R. Yakimova. Radiation hardness of wide-gap semiconductors (using the example of silicon carbide), Semiconductors, 36, (2002) 1270 http://dx.doi.org/10.1134/1.1521229

[101] J.W.Corbett, J.C.Bourgein. In Point Defects in Solids (N.Y., London, Plenium Press, 1975) v.2, p.1 http://dx.doi.org/10.1007/978-1-4684-0904-8_1

[102] Z.Zolnai,N.T.Son, C.Hallin, and E.Janzen, Annealing behavior of the carbon vacancy in electron-irradiated 4H-SiC, J Appl.Phys, 96(4) (2004) 2406 http://dx.doi.org/10.1063/1.1771472

[103] J.M.Luo, Z.Q.Zhong, M.Gong, S.Fung and C.C.Ling, Isochronal annealing study of low energy electron irradiated Al-doped p-type 6H silicon carbide with deep level transient spectroscopy J. Appl.Phys. 105(2009) 063711. http://dx.doi.org/10.1063/1.3087757

[104] G. Alferi, E. V. Monakhov, B. G. Svensson and A. Hallen, Defect energy levels in hydrogen-implanted and electron-irradiated n-type 4H silicon carbide, J.Appl.Phys 98 (2005) 113524 http://dx.doi.org/10.1063/1.2139831

[105] M. Weider, T. Frank, G. Pensl, A. Kawasuso, H, Itoh, R. Krause-Rehberg, Formation and annihilation of intrinsic-related defect centers in high energy electron-irradiated or ion-implanted 4H- and 6H-silicon carbide, Physica B 308-310 (2001) 633. http://dx.doi.org/10.1016/S0921-4526(01)00772-4

[106] G.Alferi, E.V.Monakhov, B.G.Svensson, M.K.Linnarsson, Annealing behavior between room temperature and 2000?°C of deep level defects in electron-irradiated n-type 4H silicon carbide, J.Appl.Phys. 98, (2005) 043518 http://dx.doi.org/10.1063/1.2009816

[107] A. A. Lebedev and V. V. Kozlovski, Comparison of the Radiation Hardness of Silicon and Silicon Carbide, Semiconductors, 2014, 48, pp. 1293-1295. http://dx.doi.org/10.1134/S1063782614100170

Keywords

About the Author

Professor A.A. Lebedev

Lebedev Alexander Alexandrovich was born March 10th, 1959 in St. Petersburg (Leningrad). He graduated from the Optolectronics Department of St. Petersburg Electrical Engineering University in 1983. Since 1983 he is working at the Ioffe Institute. Since 1999 he holds the position of Head of Laboratory «Physics of the semiconductor Devices». Since 2014 he is also the Director of the Solid State Electronic Division. He is a member of the Scientific Council of the Solid State Electronic Division of the Ioffe Institute. He is also a member of the Scientific Council of the Ioffe Institute and St Petersburg Electrical Engineering University.

Since 2004 he began teaching students at the St. Petersburg Electrical Engineering University and has the position of a full Professor.

A. A. Lebedev is a specialist in the field of physics, technology and device applications of wide bandgap semiconductors. He was the first who studied properties of deep centers and their participation in recombination processes in silicon carbide (SiC) polytypes. Within this work A. A. Lebedev developed a method theory and was the first who applied the non-stationary capacity spectroscopy method for the study of wide bandgap semiconductors. The results obtained by A. A. Lebedev were used for the development of the technology of SiC epitaxial layers with determined parameters. This allowed creation of experimental samples for a number of semiconductor devices (including microwave) on silicon carbide. The investigations, carried out by A. A. Lebedev and his collaborators, showed that the developed SiC devices by their limiting operation temperatures, specific commutatable capacities and radiation resistance correspond to earlier theoretical estimates. Thus, it was experimentally proven that semiconductor devices can operate at temperatures $> 800\,°C$ and radiation levels on average twice higher than limiting values for Si devices with the same operation parameters. At present the results of these investigations are introduced at Svetlana-Electronpribor, JSC (St.-Petersburg). The results obtained by A. A. Lebedev in many respects stimulated the existing world interest in silicon carbide.

A. A. Lebedev made a considerable contribution to the development of the technology of gallium nitride (GaN) and its solid solutions. Under his guidance the chloride-hydride epitaxy method was improved, which gave a possibility to grow quality GaN and AlGaN epitaxial layers by this method, including p-type, p-GaN/n-GaN homojunctions, GaN\AlGaN and GaN (AlGaN) \ SiC heterojunctions. On basis of these structures "solar blind" AlGaN UV photodetectors were obtained and experimental samples of heterobipolar n-GaN/p-SiC\ n-SiC transistors were manufactured.

Several times he received Prizes for the "best paper of the year" from the Ioffe Institute Scientific Council. Russian State medal "In memory of 300 years St. Petersburg anniversary" He published about 250 papers in Russian and International journals and International Conference Proceedings.

A. A. Lebedev was supervisor of 5 PhD students. He has four patents in solid state electronics. Dr. Lebedev's team has won and successfully fulfilled several international grant (4 INTAC, ISTC, 5 RFBR, INCO-COPERNIC and NATO SfP) and direct contracts with national and foreign semiconductors companies: (CREE (US), Schneider Electric and Thomson (France).)

A. A. Lebedev was a Member of the Program Committee of III, IV, V and VI Intern. Seminar on Silicon Carbide and Related Materials, May 2000, May 2002 and May 2004, May 2009 Novgorod the Great, Russia, Member of the Steering Committee of the European Conf. on SiC and Related Materials, Sept. 2006 (Newcastle, UK), Sept 2008 (Barcelona, Span), Sept 2010 (Oslo, Norway), Sept 2012 (Co-chairman, St. Petersburg, Russia), Sept 2016 (Greece, Member of International Advisory Board CIMTEC 2010 of Symposium H "Advances in Electrical, Magnetic and Optical Ceramics", June 2010 Tuscany, Italy; Member of Scientific Program Committee of 14-th Conference on Semiconductor and Insulating Materials (SIMC-XIV), May 2007, Fayetteville, Aransas, USA, Member of Scientific Program Committee of ICSCRM-2015, Italy, Giardini Naxos, Member of the International Advisory Committee of the International Symposium on Graphene Devices, Brisbane, Australia, 2016.

Field of research

- Epitaxy and heteroepitaxy of silicon carbide and III-N.
- Investigation properties of the SiC epitaxial films and device on it's base.
- Radiation damages in semiconductors.
- Growth and investigation of Graphene films on SiC.
- Developing prototype devices on grapheme films

Some recent publication.

1. V.V. Kozlovski, A.A. Lebedev, V.V. Emtsev, G.A. Oganesyan, Effect of the Energy of Recoil Atoms on Conductivity Compensation in Moderately Doped n-Si and n-SiC under Irradiation with MeV Electrons and Protons, Nuclear Instruments and Methods in Physics Research Section B 384 (2016) 100-105.

2. .S. Novikov, N. Lebedeva, A. Satrapinski, J. Walden, V. Davydov, A. Lebedev, Graphene based sensor for environmental monitoring of NO_2, Sensors and Actuators B: Chemical 236 (2016) 1054-1060..

3. V.M. Mikoushkin, V.V. Shnitov, A.A. Lebedev, S .P. Lebedev, S.Yu. Nikonov, T. Iakimov, O.Yu. Vilkov, R. Yakimova, Size confinement effect in graphene grown on 6H-SiC (0001) substrate, CARBON, 86 (2015) 139-145

4. V.V.Kozlovski, A.A.Lebedev, E.V.Bogdanova, Model for conductivity compensation of moderately doped n- and p-4H-SiC by high-energy electron bombardment, J. Appl.Phys., 117, 155702 (2015)

5. Lebedev A.A., B.Ya. Ber, N.V.Seredova, D.Yu Kazantsev, V.V.Kozlovski, Radiation-stimulated photoluminescence in electron irradiated 4H-SiC, Journal of Physics D: Applied Physics, 48, 485106 (2015)

6. A. A. Lebedev, E. V. Bogdanova, M. V .Grigor'eva, S. P. Lebedev, V. V. Kozlovski. Irradiation and Annealing of p-type silicon carbide, AIP Conf. Proc. 1583 (2014) 156-160

7. A.A. Lebedev, M.V. Zamoryanskaya, S. Yu. Davydov, D.A. Kirilenko, S.P. Lebedev, L.M. Sorokin, D.B. Shustov and M.P. Scheglov, A Study of the Intermediate Layer in 3C-SiC/6H-SiC Heterostuctures, Journal of Crystal Growth 396 (2014) 100-103.

8. A.M. Ivanov, N.B. Strokan, A.A. Lebedev, Radiation hardness of a wide-band gap material by the example of SiC nuclear radiation detectors, Nuclear instruments and Methods in Physics Research A, A 675 (2012) 20-23.

www.ingramcontent.com/pod-product-compliance
Lightning Source LLC
Chambersburg PA
CBHW071642210326
41597CB00017B/2079